Do you know enough to care?

The
Energy
Trail

. . . Where it is Leading

An essay on what we can expect in the not too distant future

George H Croy

Society of Petroleum Engineers (SPE)
South East Asia Petroleum Exploration Society (SEAPEX)
Technical Journal Editor (retired) & Freelance Writer

 **World Scientific**

NEW JERSEY · LONDON · SINGAPORE · BEIJING · SHANGHAI · HONG KONG · TAIPEI · CHENNAI

Published by

World Scientific Publishing Co. Pte. Ltd.

5 Toh Tuck Link, Singapore 596224

USA office: 27 Warren Street, Suite 401-402, Hackensack, NJ 07601

UK office: 57 Shelton Street, Covent Garden, London WC2H 9HE

British Library Cataloguing-in-Publication Data
A catalogue record for this book is available from the British Library.

THE ENERGY TRAIL — WHERE IT IS LEADING
Do You Know Enough to Care?

ISBN-13 978-981-281-857-7 (pbk)
ISBN-10 981-281-857-X (pbk)

Design by **LAB Creations** (A Unit of World Scientific Publishing Co. Pte. Ltd.)

Printed in Singapore.

Reviews

Have no… (*happy ever after*)… preconceptions about this book - it has a 'doom and gloom' ending! It somewhat brutally leaves the reader without solution - but this is its strength! Agree or disagree, like or hate the ending, you will be compelled to acknowledge the message and be provoked into thinking about the ethical use of energy.

This book makes it clear that we have already 'lost the plot' with our global energy use - and only the stark reality of such radical thoughts will… enable us to arrest the otherwise inevitable demise of society as we know it! - *Alan J S Lilley, Managing Director, Vision International Limited*

A subject of global importance, (*the author*) presents it in relatively few words, understandable by the average person but with sufficient scientific content to (*be of interest*) to others (*of a higher persuasion*)… - *John Westwood, Managing Director Douglas-Westwood and acknowledged expert on global hydrocarbon reserves*

The book maps the trail of our dependence on the hydrocarbon molecule, without casting blame on the 'evil' oil industry. That is a breath of fresh air… - *Richard A. Lorentz, VP Business Development, Aabar Petroleum and Chairman of Franklin Offshore Marine*

As with most text on this subject, authors do not write the book unless they have a belief in one side of the argument or the other. (*The author's*) position is that the current trends in energy consumption and environmental fallout are untenable. (*The author*)… has an interesting conversational style and is quite accomplished at describing fairly complex chemical and physical concepts in terms that most people will understand… - *Phil Rae, SPE Distinguished Lecturer and Director, InTuition Energy Associates*

The Energy Trail...

... Where it is Leading

Introduction

When I started writing this book several years ago, my intention was to produce a relatively easily read book that would guide the reader through our quest for energy and what it has done to us, to our planet. It was necessary to introduce some basic physics and chemistry, which I have gone to great lengths to make as understandable as possible, for I believe that one of the reasons we are in the situation we find ourselves today is lack of knowledge. How can a relatively intelligent person understand what is happening around him without first having some knowledge of his surroundings and how they affect him and he affect them? As I have said to my children through their formative years and into their adulthood, knowledge alone is a dangerous thing, without the intelligence to interpret that information and understand how it applies to you and your life. Knowledge and intelligence go hand in hand.

I still remember, when I was about seven years old, in Burnside, Rutherglen, just outside of Glasgow, walking down High Burnside Road and seeing this huge, blue sun. Blue sun, not a blue moon. These things happen once in a blue moon, but this was a blue sun. And it was huge. Much bigger than the sun normally would have been. I wasn't scared, I was curious, but it wasn't until many years later that I finally found the answer - that a cloud of gas from Colville's Steel Mills in Motherwell had passed between me and the sun, causing a distortion in the light, making the sun look both bigger and blue. As I remember, it wasn't a very momentous occasion, but it did instil an insatiable curiosity in me to find out everything I could, about everything. I never took things for granted. I never accepted without question explanations from grown ups. I also took to reading ravenously any books, but especially books on science - chemistry, physics, astronomy, geology, biology etc., and I became well known for breaking off a meal, should an argument arise, to go and find a book to verify my side of the argument!

Too many people today have access to information without having that knowledge-intelligence combination needed to make sense of that information to their own benefit or to understand what it is they are doing to this planet. A housewife steps into the kitchen and switches on the stove to cook dinner. She never questions where the gas or electricity came from to provide her with that facility, nor does she worry, or even understand that she is contributing to the global atmospheric heating that is causing so much concern. But if you tell her that two billion people around the world are doing exactly the same thing as her, she surely has the intelligence to begin to understand how big an impact that could have on our global temperature.

If you open your refrigerator, you never pause for a second to marvel at what an ingenious piece of machinery it is, or to wonder where all the food produce within it came from. But just imagine a billion people, a billion refrigerators, all full of food - how did we get there, where did it come from? I doubt if there are many of you, reading this book, who can comprehend even what a million looks like, let alone a billion. I know I have difficulty with it. Carl Sagan, a very famous scientist and author, claimed that he never used the term 'billions of billions' on a Johnny Carson Show, as it was incomprehensible and far too vague a number for the human mind to comprehend but he eventually surrendered to the need to use such numbers, in what turned out to be his last book of an illustrious career in science and writing, 'Billions and Billions'. I recommend you find it, and read it. Just taking a few comments from his book, there are 60 seconds in one minute, and a million seconds in 12 days. A billion, is 1000 times that, 12,000 days or 32.8 years! So, in these terms, six billion people are equal to 197 years if each one represented one second. Each one needs to be fed, each one needs energy. Where is it going to come from?

As this book progressed, as I delved deeper and deeper into the ever changing conditions in which we live, I began to realise that the situation is not getting any better, that it is in fact getting considerably worse. A report, started in 2001 at the initiative of the UN and involving some 2000 authors, reviewers and researchers around the globe who contributed their knowledge, creativity, time and enthusiasm, was launched in early 2005 — the Millennium Ecosystem Assessment. While it shows a real attempt at understanding what we are doing to this planet and hopefully will get the governments of the world to act together to try and stem the destruction of our home, it is the people of the world, people like you and I, who need to act, to act decisively, to try and prevent the utter destruction of the human race.

We are a product of this planet. We have an inbuilt belief that the planet will look after us, will care for us. I hope, as you read through this book, that I can dispel that belief and get you to understand that your survival, your children's and your children's children's survival depends a great deal on what you do tomorrow, and every day for the rest of your life.

In this book I pull no punches. Some people will ridicule the book. They will say "…he doesn't know what he is talking about". My response to these people is to ask them, "…do you have the time to wait and see?" Take a look around you. We have enough historical data to know that things are changing, changing dramatically. Every day the News confirms what I'm trying to put across - power shortages, Government concerns about the future, higher costs, etc.

The greatest misfortune of the planet is that we evolved into an intelligent race of creatures. But there was so much we did not understand! There was so many things to be afraid of because lack of understanding brings fear. We discovered religion. Religion became the panacea for all things not understood — the sun, fire, death — we created a God, or gods to take responsibility for all these things. But religion also brought with it power, the power of control, the power of fear. Man was created in the image of God, so how could he do any wrong? Our lives were controlled by religion, not by our own intelligence. We knew nothing beyond our own small gathering, a gathering that was controlled by religion.

And therein lies the biggest hurdle to our survival as a human race. So long as we continue to believe that we have a God-given right to this planet and its resources, so long as we continue to see ourselves as superior to all other living species, we will continue to dig our own graves. I am not talking about our ability to grow food, provide ourselves with a few creature comforts, I'm talking about our ability to evolve further, to live on this planet as an advanced, intelligent form of life that can go onto greater and higher things — the eradication of poverty, sanitation for everyone, full education for the masses - the list can go on for ever, the things we could do, the things we should be doing, instead of satisfying our own short-term needs. At our present rate of destruction, within less than a thousand years we will be relegated to living in small isolated cities - townships, with no industry worth talking about and probably very little food, save what can be grown in greenhouses. Because that will be the only form of life left to us with the power generation systems we will be able to sustain. Smaller pockets of humanity, with no access to power other than that provided by the Sun, if we can still see it through the pollution, will rapidly die out.

If we leave things unchanged, if we continue to live as we have done for the past century and more, using up our energy heritage as if it were infinite, expanding the global population (10 billion by 2050?) each with his or her own demands on the systems that support us, we will run into a brick wall. In fact, some scientists suggest that we have already reached that brick wall. Having done the amount of research I have for this book, having looked at situations from both sides of the equation I have to agree with those scientists.

That is what this book is about. It is about trying to warn people before it is too late, it is about where this energy trail is leading us. I hope that people will read it with an open mind and get some understanding of this merry-go-round that we are on, for which we only have ourselves to blame. I hope that they will begin to understand the future we are providing for our children, for our grandchildren and unborn generations to come. We don't need nuclear war to destroy the human race - we are doing it already, of our own free will.

George H Croy
Singapore
January 2008

Dedication

I begin by dedicating this book to my mentors, the authors who raised and nurtured the curiosity in me to dig deeper, to search further — to people like Arthur C. Clarke, Jagit Singh, E. Sheldon Smith, Fred Whipple, Carl Sagan, Isaac Asimov, George Gamow, Charles Darwin, Will Durant, Erwin Schrödinger, anyone who ever wrote a book on science… the list is endless, but how sad our world would be if it were not for people like them who have knowledge to share. If only we could encourage more young people to develop the insatiable curiosity that these authors created in me, the world might yet be saved.

I also dedicate the book to my four wonderful daughters — Tracy, Amanda, Samantha and Kimberley, my TASK force — that drove me to write this book, that they may understand, with sincere and deep regrets and apologies for the world I will be leaving them with. If we had only known sooner and started doing something about the situation while we still could. I hope that they and their generation will be able to at least curtail if not stop altogether the destruction, so that my grandchildren, and their grandchildren might have a future.

But I dedicate especially this book to Chew Siok Kim, my dear and darling wife of more than thirty years, without whom life would have been so much less meaningful. Perhaps some day she will understand how much she has meant to me all these years.

Contents

The Energy Trail...

"Life on Earth is driven by energy.

Autotrophs* take it from solar radiation and heterotrophs[†] take it from autotrophs. Energy captured slowly by photosynthesis is stored up, and as denser reservoirs of energy have come into being over the course of Earth's history, heterotrophs that could use more energy evolved to exploit them, Homo Sapiens is such a heterotroph; indeed, the ability to use energy extrasomatically[‡] enables human beings to use far more energy than any other heterotroph that has ever evolved. The control, of fire and the exploitation of fossil fuels have made it possible for Homo Sapiens to release, in a short time, vast amounts of energy that accumulated long before the species appeared."

Dr David Price, Energy and Human Evolution[1]

*An organism capable of synthesizing its own food from inorganic substances, using light or chemical energy. Green plants, algae, and certain bacteria are autotrophs.

[†]An organism that cannot synthesise its own food and is dependent on complex organic substances for nutrition, e.g., humans.

[‡]Outside the body.

01

AN INTRODUCTION

Energy has been with us since time began, even before the first ray of light shone down on the proto-planet that was to become Earth. In fact, light is in itself a form of energy.

Without it - ENERGY - we would not exist...

Without it, we would never have evolved and developed to the place we hold on this planet in which we today find ourselves. But what place, what position is that? The answer is the basis of this book.

I do not presume to present myself as an expert on any of the topics herein. There are those who have much higher claim to that title. However, I have been involved in the energy industry for more than thirty five years in one way or the other and it is that experience and accumulation of knowledge I hope to share with you.

Although subdivided into chapters, this book is broken down into three basic sections:

■ **How we got here**

■ **What we are doing here**

■ **Where we go from here**

The primary source of all energy, for our small planet at least, is the Sun. The known universe is full of suns, but our Sun is the only one close enough to provide sufficient energy to allow Earth to have developed as it has.

But what is energy? Put one way, energy is the single most important technological challenge facing humanity today. Nothing else in science or technology comes close in comparison according to Professor of Chemistry at Caltech, Nate Lewis PhD (MIT).

Or, put as simply as possible, energy is a 'store', a repository of an ability to do work. You cannot 'see' energy, it is a concept. What you see are manifestations of energy transfer, energy conversion or the energy 'store'. When someone says they are awed by the 'raw energy' of a raging sea, what they really are witnessing is the expenditure of the energy that drives the system, into something else. Similarly, a battery is not 'energy', it is an energy 'store' - it has the 'potential' to provide energy, given the right conditions.

Energy, by definition, now amply supported by experiment, cannot be destroyed. It can only be converted into some other form of energy. Anyone driving a powerful, gas-guzzling six or eight cylinder-engined car would probably beg to differ, but in fact that is only because they have not taken into consideration the total dispersion of the energy inherent in their litre of gasoline.

That litre of gasoline will release a certain amount of energy, energy that is inherent in the bonding of the atoms that constitute the liquid. Imagine if you took a litre of gasoline and sealed it into a solidly built container, provide oxygen and an ignition source and watched what happens. There would be an explosion sufficient to take the windows out of every building in the vicinity. This is because the litre of gasoline has been converted into other forms of energy - light, sound, heat, etc.

Basically the same thing is happening within your car engine cylinder - light, sound, heat. The light is wasted energy, it has nowhere to go. The sound is also wasted energy. In fact, most cars now have insulation around the engine compartment to try to muffle the sound, as it is more annoying than useful. The only time you are likely to hear it is when your engine misfires and you get a loud 'backfire' sound.

So, out of your litre of gasoline, all you have acquired that is useful is heat. This heat expands the gases created, driving the cylinder downwards, which in turn rotates a crank, converting heat into rotation, which is eventually converted into distance, as your car moves.

Some of this may be Double Dutch to you, but please bear with me, all will be explained, eventually. What I'm endeavouring to do right now is to show how we have learned how to waste energy.

You have converted liquid (chemical) energy - petroleum, or gasoline - into distance. But at what cost? So far we have 'lost' energy to sound and light. The heat generated, while used to expand the gases in the cylinders of your engine, also heats up the engine block, which would eventually burn up (as anyone who has run out of water in their engine radiator will know). This heat is eventually vented into the atmosphere through your engine's radiator.

An internal combustion engine is in fact an extremely poor converter of energy. A figure of about 25% efficiency is considered average for a well-maintained engine. This means that 75% of the energy contained in that litre of gasoline goes out the window! Or rather, up the exhaust pipe. And that's in a well-maintained engine! How much less efficient are the rest of the vehicles on the road today?

The above was not, is not, intended to be a lesson on the internal combustion engine, nor is it a lesson on energy conversion. The sole intention is to, I hope, highlight how inefficient in many ways is our technology and how much we have wasted. Only now, are we beginning to investigate this wastage and the effects it is having on our environment.

Anyway, back to our energy quest. Energy, as stated above, cannot be destroyed, only converted from one form to another. A simple illustration would be to lift a normal, red clay, kiln-fired building brick up on top of a wall. That brick now has what is called potential energy, i.e., a potential to supply energy by virtue of its position - on top of the wall. But, it took energy to get that brick up there in the first place, didn't it?

Where did that energy come from?

In a roundabout way, it came from the Sun. You expended energy in lifting the brick up onto the wall, in this case the chemical energy of your body (glucose), which you converted into mechanical energy through your muscles. That glucose energy in turn came from the food you eat, which was itself, in turn, 'energised' by the Sun during its growth.

Fairly simple so far, isn't it?

Something a bit more complicated however, is the equation, $E = mc^2$, where E equals energy, m is the mass of the material in question and c is the speed of light in a vacuum, an equation proposed by Albert Einstein in the earlier part of the 20th century but only proven, with the detonation of a device at the Trinity Site at Alamogordo in New Mexico many years thereafter - the 'Atom' bomb. Since E = energy (in electron volts), just think of the quantity released by the detonation of an atomic bomb, where m = mass (in kg) and c = the speed of light in a vacuum – 300,000,000 m/sec.

Mass, (anything having weight), could be changed into energy!

Man finally found an alternative to the Sun in energy production. Even so, the Sun did have a lot to do with it. Our Sun, or at least its kind, had everything to do with it actually, but not the subject of this present argument, which is the transformation of energy from one type to another. We will come back to atomic, or nuclear energy later.

Energy comes in many different forms, but four types will suffice for most arguments. Remember, energy is the 'potential' to do work. The brick on top of the wall has potential energy, as does water in a reservoir, poised above a turbine generator. The litre of gasoline has energy, this time it is chemical energy, inherent in the bonds of the atoms and molecules of which it is made (explained later).

Kinetic energy, (from the Greek *kinetikos* - move) is the energy inherent in an object by virtue of its movement. Imagine a hammer descending rapidly towards a nail head. The hammer possesses 'kinetic' energy. (So does a car travelling at 150 km/hr down the expressway. Lots of it).

The fourth type of energy of interest to us in this book is nuclear energy. Science divides energy up into many other recognised forms, internal, heat, light, etc., but the main four mentioned are forms from which most others are derived.

For millennia, Mankind was satisfied with the most primitive of energy sources - fire - a natural phenomenon he learned to control and master. This fire, he made by burning flammable material - grasses, wood bark, twigs and fallen branches. He used fire primarily for light - it scared away animals - and heat, allowing him to move further north on the global continental landmass. Only later did he find out that cooked meat tasted better than raw, using the fire to roast animals he had slaughtered (and thereby laying the groundwork for one of the most dreadful diseases to which Man in particular is prone – stomach cancer – caused by the energetic alteration of protein into cancer-causing agents, i.e., burning it!)

However, the stuff that dreams are made of, and wars fought over, the stuff that drove the most rapid expansion of mankind in the history of our time on this planet, arrived on the scene probably millions of years before that, long before Man, except that Man, when he arrived, did not recognise its potential. Not until much later.

Before we go there however, let me digress a moment to define one or two of the more important forms of energy, most notably 'heat'. It is almost inconceivable to think that people are not familiar with heat. You cook your meals using heat. You warm your body using heat. Heat can do a multitude of things that we all find beneficial. But what is heat?

Heat, as we've already decided, is a form of energy. Temperature and heat have only a passing acquaintance although most people consider heat and temperature to be the same thing. A chunk of ice contains less heat than a pan of boiling water, their temperatures vary by a 100 degrees Celsius, or Centigrade, (180 degrees Fahrenheit) so heat and temperature are synonymous – yes, but not quite. Temperature is a measure of differences in heat content. Differences in heat content are an example of differences in energy content. But then, when energy is continued to be applied to water at its boiling point, it will continue to absorb that energy with no increase in temperature. Similarly, if you continue to extract energy from water at zero degrees Celsius, it will not change in temperature.

The heat absorbed (or extracted) without temperature change is known as latent energy. This is the energy of destroying (or creating) the bonds between the molecules of water, to form steam (or ice). Most people don't know what steam is, because it is not the white fluffy cloud you see coming out of your kettle – that is water droplets made from condensed steam – real steam can be seen (or rather, not seen!) in the very narrow band of clear air at the spout of the kettle. Touching that very narrow band will give you a very severe burn, much worse than if you touch the 'steam' of the white cloud. Any idea why?

The reason is that the clear air section is composed of completely disassociated water molecules. It holds not only the heat of boiling water, but also the heat of breaking the weak bonds between the molecules, the latent heat. The moment if touches your skin, the latent heat is dumped as the water molecules recombine to create water droplets. Of course, your skin absorbs the heat, a lot of it.

All atoms and molecules are in motion. This was first discovered by a Scottish botanist, Robert Brown, around 1827 while inspecting pollen under a microscope. His discovery led to the conclusion that all molecules (and therefore atoms) were in motion – the higher the heat content, the faster the movement. And so it is, throughout the physical world. Everything (everything!) is in motion courtesy of its quality and heat content. Or put another way, heat content is a measure of the state of agitation of the constituent molecules or atoms. Imagine a piece of steel. At room temperature, it is a solid, inflexible piece of metal. Add some heat to it and it will slowly become 'malleable' as in the material of the sword maker's blade or the blacksmith's horseshoes. Add more heat to it, and it becomes a liquid, which can be shaped and designed according to specially made moulds. Add even more heat to it, and it will become a gas.

Notice the progression? Solid, liquid, gas. Each phase is a step change in energy content, or heat. The more heat applied, the more agitated the molecules or atoms become, first they become plastic, then flowing before finally breaking free from each other as a gas. Each phase change also brings with it an increase in volume – something we will come across frequently as we talk about energy.

A couple of final points before we move on. And this is where temperature comes back into the picture. As mentioned, temperature is a measure of differences between heat content. So, if we draw a graph of temperature vs. heat content for any substance, we end up with a pretty straight line. And this was put to significant use by William Thompson, later Lord Kelvin, when he discovered what became know as zero degrees Kelvin, that point at which there was no heat, therefore no motion – a point at which no material substance can exist. Hydrogen, which becomes a metal when it freezes, does so around 17 degrees Kelvin.

In fact, the freezing, (or boiling) point of each and every atom or molecule is a well defined and known temperature today, and it is used to separate them from each other, especially gases.

The second point is that 'temperature' is not a measure of ALL the heat in a solid, liquid or gas. But more of that later.

On with the quest!

■ How we got here

Hydrocarbons, or fossil fuels*, are the remains of organic material

---

* 'Fossil fuels' is actually a misnomer in the sense that a fossil is the petrified remains of a once living organism, i.e., a chemical replacement process, such as the petrified forests of Arizona USA and Greece, where the soft tissues of plants and trees have been slowly replaced by chemicals to form rock. Hydrocarbons were created by anaerobic decay.

(plankton, etc., in the case of oil and gas; trees and ferns, in the case of coal) and were formed three, four hundred million years ago, during an exceptionally suitable climatic period, warmer than today, when all forms of living material proliferated. The death of these creatures, and plants, their subsequent sedimentation under sometimes kilometres of soil and the heat generated at those depths, in anaerobic conditions, which allowed them to decompose in the absence of oxygen, produced oil, natural gas and coal. Over subsequent centuries, millennia, aeons, some of the liquid and gaseous products of this process migrated back to the surface, through fissures in the rock, through earthquake action, to appear as oil seeps, pockets of asphalt or bitumen, and gas flares.

There are records in ancient scripts referring to the uses Man first made of this unusual material - caulking boats, mummifying bodies, artistic decorations - many things, except as a source of energy! Just imagine how the world might have developed if they had!

Evidence of Man's occupation of the lands between the Mediterranean and the Caspian Seas dates back more than 40,000 years. Known as Mesopotamia, part of the Fertile Crescent, which stretched from the Nile Valley in Egypt to the southern tip of today's Iraq, it lay between the rivers Euphrates and Tigris, and became the home of civilisation.

Here, Man first learned cultivation and the beginnings of a civilised life. In those earliest of times, pools of hydrocarbon liquids were undoubtedly found, and constant fires from gas leaks were a hazard of everyday life. In fact, information from cuneiform tablets, found in the ruins of ancient Babylonian cities, attests to the presence of massive oil and gas induced fires, creating a sense of awe and helplessness in the populace: "...if in a certain place of the land naptu oozes out, that country will walk in widowhood..." and again, "...if a pit opens in fallow land and burning naptu appears, the land will be destroyed..."[2]

Naptu, is the root of the Arabic word Naft, meaning crude oil and the Greek Nafta, from which we get our English word Naphtha, a principal and important part of the hydrocarbon chain. Again, more later.

By five thousand years ago, this Fertile Crescent area was home to a sophisticated civilisation, with cleverly irrigated fields, domesticated animals and large city-states such as Sumer, Babylon, etc. In these places, hydrocarbon fractions found their way into multiple uses, including oil lamps and for medication, both internal and external. Heavier fractions of the hydrocarbon family were used in a multitude of tasks. Kupru, or bitumen found an ideal use as a waterproofing agent as well as a glue or cement since, by its sticky nature, it stuck to almost everything. It was used in boat building (the basket in which Moses, 3300 years ago, is supposed to have been placed on the Nile was waterproofed with bitumen*), mummification of dead relatives, in art and jewellery - it was even used to make sandals. In fact, evidence exists from archaeological digs that the use of bitumen as a tool, for holding weapons together, dates back as far as 38,000 years ago.

So Man's involvement with hydrocarbons goes back many, many centuries, many millennia, with little evidence that he realised that this was a massive store of energy just waiting to be tapped.[†]

Involvement there was however, in barter and trade and much of the history of the Near East, Egypt, Greece and Rome revolves around this product. It was traded along the Silk Road, both as a glue and as a fuel – they burned lumps of tar. Feuds erupted and city-states were destroyed over the possession of this material, even as they do still today. Except,

---

*This story is remarkably similar to that of King Sargon of Assyria, recorded 2300 years earlier. (As a follow-on to that, the burning bush that Moses is supposed to have seen, was in all probability a gas vent that caught fire. Just imagine what could have happened if his curiosity rather then his fear had gotten the better of him, and he had discovered the power of hydrocarbon. The Jews would have had the oil instead of the Arabs. This is similar to the Jewish lament that if Moses had only turned right on emerging from the Red Sea, instead of turning left.)

[†]This has to be looked at from the context that around 10,000 years ago, the global population was something like five million. By Anno Domini it was 200 to 300 million and by mid 17th century it was 500 million. In 1800 it was one billion and it only took 130 years for the population of the world to reach two billion. It took 30 years from there to reach three billion; and by 1975 it was four billion; in 11 more years it was five billion. It topped six billion before the end of 1999 and is estimated to reach over nine billion before 2050. Most of the growth can be attributed to the availability of energy – the Hydrocarbon Era. Read on and decide if you would have wanted our ancestors to have developed hydrocarbons any sooner.

the worth of the hydrocarbon was not as an energy source - as it is today - but for the purposes and reasons given above.

By the six or seventh century it had already found use as a weapon, being used as a fire ball to set fire to opponents' ships, buildings, etc. Launched as a fiery missile from catapults, it burned wherever it landed, even on water. Then, somewhere around the 15th century, there is evidence that people of the region were practising some form of refining, suggesting that they were beginning to understand the potential uses of this material. It was refined to produce oil for lamps.

At this time, all the hydrocarbon materials being traded, marketed, stolen - fought over - were taken from surface exposure - oil seeps, oil lakes, pits, etc., - or dug up from deposits such as tar sands. But Man's need for this product was growing by leaps and bounds. Empires were being made and destroyed over its possession; trade was flourishing from all corners of the known world. More of this stuff had to be found and it was the activities that followed that really mark the beginning of what is colloquially known as the Hydrocarbon Age. One event, which encouraged Man's entry into the Hydrocarbon Age but which, ostensibly, had no bearing on it, was the need to lift water for irrigation.

The name Huygens probably means little to anyone who doesn't wear spectacles, (in fact, it probably means very little to those that do!), but Dutchman Christiaan Huygens, apart from being a brilliant physicist who discovered the light wave effect and did a lot of work on lenses, in 1680 devised a means to mechanically lift water, by exploding small charges of gunpowder in a closed cylinder, so as to drive down a piston - the basis of today's internal combustion engine. Of course, it didn't work, but the principle was put in place. Part of the reason it did not work was the crudeness of the mechanical parts and the amount of energy sacrificed in trying to overcome the poor quality. Remember, energy cannot be created or destroyed.

Many people must have fiddled with the idea of using hydrocarbons as a means to drive an engine, but it wasn't until almost two hundred years after Huygen's experiments that a French engineer, Jean-Joseph Étienne Lenoir, in 1859 managed to convert a double acting steam engine into an internal combustion engine. By adding a spark ignition to the cylinders and, instead of steam, injecting a combustible mixture of coal gas and air, Lenoir created the first, continuous operation, internal combustion engine. Mind you, it was all of four percent efficient, most of the energy disappearing into the wide blue yonder as sound, heat and unburned fuel. But it set a precedent.

A company in North America, the Pennsylvania Rock Oil Company, that had made its money from rendering whale fat to make lamp oil, in the 1850s figured that digging this rock oil out was both time consuming and expensive[1] and developed a method, with the help of a certain gentleman, 'Colonel' Edwin L. Drake, of drilling a hole to extract the oil. The 'Colonel' was a title he gave himself, he being only a railroad conductor, as he felt it increases his stature with the media. The method he used was an adaptation of the salt mine drilling equipment which had already been in use for more than half a century.[2]

Using steel-faced chipping tools and heavier piping, the percussion drilling which became the modus operandi for oil well drilling for the next forty or fifty years, in 1859 brought about the first oil 'discovery' at Titusville in Pennsylvania, when Drake dug through into a cavern at 69 feet, got stuck, and oil floated to the surface of the water that filled the hole.

The Czar of Russia, hearing of this discovery and the potential it had to change the balance of power in the oil supply market, began the earnest development of the oil fields around Baku, on the coast of the Caspian Sea, resulting in the world's first blowout, when one of the wells struck high pressure oil.

Most notable amongst the investors that set up business in Baku were brothers Robert and Ludwig Nobel who, in 1873, set up the establishment of the Nobel Brothers Oil Extracting Partnership.

Then in 1901, in Spindletop, Texas, Anthony Lucas, using a rotary drilling rig, created not the first, but definitely the most famous American blowout, from a 300m deep hole, producing 80,000 barrels of oil per day, an unheard of amount from an oilwell, until that moment.

The world has never been the same since then.

Although the Egyptians are credited with using rotary drilling mechanisms to drill for water as much as 5000 years ago, it was not until much later, in 1500 that Leonardo DaVinci developed a design for a drilling rig that is similar to many of today's rigs. Today, most of the wells drilled use conventional rotary drilling rigs to dig their deep wells, although advanced technology is also changing that.

The more recent developments of the rotary drilling rig were without question the turning point in Man's utilisation of hydrocarbon as an energy source. Until then, oilwell drilling had been conducted by using a percussion type system, entailing dropping heavy, steel-faced tools into the hole, on the end of a cable or rope, so as to shatter the rock. The broken rock then had to be bailed out using a bucket.

This method was very limiting, both in time and in depth. One of the problems faced was the continuous incursion of water, from surface aquifers, into the well. Then someone came up with the idea of using this water as a means of removing the rock cuttings, as it already was doing, even in their bailing operations, by pumping a mud-like mixture down the pipe and up the outside, while rotating the pipe. This method extended well drilling capability from a few hundred metres to many thousands of metres. The oil industry as we know it today was borne.

Before getting too deeply involved in the development of the hydrocarbon energy industry, it will help to get a better understanding of this product and its potential.

As mentioned before, hydrocarbons are produced by the decay and decomposition of once-living organisms that have been buried in anaerobic circumstances (i.e., an oxygen-free environment).

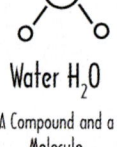

Hydrogen (atom)

Oxygen, O$_2$

Both are Molecules

Hydrogen, H$_2$

Water H$_2$O

A Compound and a
Molecule

The principle element in a hydrocarbon molecule is carbon, denoted by the letter 'C'. Another component is hydrogen, denoted by 'H', giving the name hydro-carbon. There are no other elements involved in pure hydrocarbons, but the abilities of these two elements to be attached to other elements make hydrocarbon molecules suitable for a vast array of compounds. (An element is composed of a single type of atom, a compound is composed of two or more elements – O$_2$, two oxygen atoms combined in one molecule, is an element, CH$_4$, methane, a molecule of one carbon and four hydrogen atoms, is a compound, as is H$_2$O, water).

Hydrogen is the simplest atom, with a single proton (positively charged elementary particle) surrounded by one electron (negatively charged elementary particle). The reason there is always two hydrogen atoms in naturally occurring hydrogen is that electrons are most stable when they inhabit what is known as a full shell or 'cloud' around the two hydrogen atoms. Single hydrogen atoms with one electron do not exist in nature for more than a brief moment in time, as they are highly reactive and will react with virtually anything. These 'clouds' surround all atoms and exist in various 'stable' states, or layers, according to the number of protons and the number of electrons, in a series that goes 2, 8, 18, 32, etc., one electron for each proton. (See Appendices IV, V)

Again, it is not the intention to teach basic chemistry, but what is presented here will help to understand the process of the utilisation of the energy inherent in hydrocarbons. Suffice to say that these 'islands' of stability have a lot to do with energy. Here, we are looking at the ionisation potential*

*The ability to change a stable atom into an ion, a charged particle, by the removal or addition of an electron to that atom.

of the individual electrons. Electrons in the outer stable shell are very hard to remove from the atom, i.e., the attraction of the positive proton to the negative electron is very high - it takes a lot of energy to separate them.

However, not all atoms can have stable outer shells. Let's see, two (2), would be helium. Helium is a very stable, single atom, unreactive gas. What has eight (8) electrons in the outer shell? It must have two plus eight protons - ten (10) - or neon, another stable, single atom, unreactive gas. And so it goes on. All the rest of the atoms are 'unstable' and usually are perfectly willing to 'share' electrons with another atom to reach a stable configuration. Those with a few electrons more than a complete shell will share one, or two electrons, while those with an almost full shell will willingly grab an electron offered for sharing. Such 'sharing' is referred to as a 'bond'. Some 'sharing' is very energetic, and is called an exothermic (giving off heat) reaction. Others, called endothermic reactions, may require energy (heat etc.,) to occur. Once created, these bonds are a store of energy.

Basic science tends to look at atomic structure as made of 'particles', discrete packages that can be identified as such. Basic science imaging shows a football-sized proton with a golf ball-sized electron whizzing around it at amazing speed. In actual fact, you can't see something like that, and it might benefit the reader more to think of electrons as 'clouds' of energy surrounding the proton, that have a very specific value, equal to, but opposite in charge to the proton. In this way, it is easier to understand how 'clouds' could merge and mix when atoms combine.*

*It might benefit the reader to get a hold of a book by Bill Bryson called 'A Short History of Nearly Everything' in which he gives very vivid descriptions of atoms, electrons, protons etc.

The classical concept of an atom was of a very dense core surrounded by electrons, not altogether incorrect, considering Ernest Rutherford's experiments that led him to his conclusions regarding the structure of the atom, "It was as if one had shot a large naval shell at a sheet of tissue paper and it had bounced back!"

The best way to imaging an atom is to think of it as a very dense core with a cloud of negative charge surrounding it. The higher the atomic number of an atom, the denser will be the electronic cloud. To all intents and purposes, atoms are about the same dimension irrespective of their atomic weight

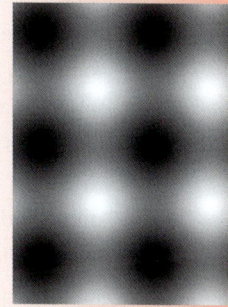

Atoms in a matrix look something similar to this, the atoms 'sharing' their negative energy clouds.

The simplest hydrocarbon is CH_4 - or methane. A single carbon atom has a valence of four (4), which means it has a strong desire to 'share' its four 'spare' spaces with other atoms to attain a stable eight (8) electrons in its outer shell. Knowing the above, a question arises. Carbon (six protons, six electrons) has an inner shell of two (2) electrons and an outer shell of four (4) electrons. Is it possible for carbon to reach a stable state (of 2) by 'lending' its four electrons? Not very likely. The energy required to separate electrons from its carbon proton nucleus would be impossible to achieve (see table in Appendix V). The carbon nucleus must 'attract' four more electrons. (One electron looks exactly the same as another electron to a nucleus – so it makes little difference where they come from!)

In chemistry, only the electrons of the outer shell participate in any reaction. And it is this desire to reach a stable shell of eight electrons that makes carbon so versatile, and such a good energy source. It will combine with oxygen, to produce CO_2 giving off energy in the process (a process we recognise as 'burning') or with another carbon atom, to produce a different compound, such as C_2H_6, ethane. Carbon atoms can link together almost ad infinitum, in massively long chains, each with its complement of two hydrogen atoms, as in $C_{30}H_{62}$ a long chain hydrocarbon, which, at room temperature, is a solid. Hydrocarbons can range from gases, (methane, ethane, etc.,) to liquids, (petroleum, kerosene, etc.,) to solids such as bitumen.

Carbon atoms can combine with one, two or three carbon-carbon links, so that the three compounds, ethane, C_2H_6, ethylene, C_2H_4 and acetylene, C_2H_2 are basically the same original molecule, C_2H_6 with one, two or three carbon bonds, minus of course the displaced hydrogen atoms. Each C-C bond makes the compound more reactive and, while ethane is flammable, acetylene is explosively so. The energy it takes to hold the atoms together, the bonds, is what is released when we 'burn' these hydrocarbons, turning the bond energy into heat energy (and light & sound). The explosive nature of acetylene is caused by the relatively unstable condition of its bonds – on Earth at least, it is a manmade product

$$H-\underset{\underset{H}{|}}{\overset{\overset{H}{|}}{C}}-H \quad \text{Methane}$$

$$H-\underset{\underset{H}{|}}{\overset{\overset{H}{|}}{C}}-\underset{\underset{H}{|}}{\overset{\overset{H}{|}}{C}}-H \quad C_2H_6 \text{ or Ethane}$$

$$C=C \quad C_2H_4 \text{ or Ethylene}$$

$$H-C\equiv C-H \quad C_2H_2 \text{ or Acetylene}$$

$$H-\underset{\underset{H}{|}}{\overset{\overset{H}{|}}{C}}-\underset{\underset{H}{|}}{\overset{\overset{H}{|}}{C}}-------\underset{\underset{H}{|}}{\overset{\overset{H}{|}}{C}}-\underset{\underset{H}{|}}{\overset{\overset{H}{|}}{C}}-H \quad C_nH_{2n+2}$$

Carbon and Hydrogen atoms can combine in many different ways, making hydrocarbons one of the most versatile compounds available to Man yet all we can think of to do with it is burn it as fuel mostly.

and not found as a gas in nature although it does exist in interstellar space and on some of the gas planets' moons.*

Because of the decisions of chemists and scientists, in the 18th and 19th century, that all carbon compounds had to come from originally living organisms (by that time scientists had figured out that all living species contained carbon in their structure), the study of carbon and carbon compounds became known as organic chemistry. The fact that subsequent experiments produced very non-organic, non-living carbon compounds did not faze the scientists, even until today, so that the study of any carbon compound is referred to as organic chemistry. This, however, included almost 95% of all known compounds, making carbon unquestionably the most versatile element in the Periodic Table, the list of all know elements[†].

The Earth is composed of various rock types, all of which, at least the ones of any interest to us in this discussion, occupy the first few kilometres of the planet's skin. These are sedimentary, metamorphic and igneous rocks. Sedimentary rocks are formed by the decomposition of existing rock by weather action and water (sandstones), or by the death of marine creatures (limestones), the product of which is washed down into sedimentary basins or shallow sea beds. As they accumulate, the overburden, or accumulated weight, drive the sedimented rock deeper and deeper into the Earth's crust, compacting the particles which, over time, become cemented together by chemical deposit from interstitial water.

If driven deep enough, the rocks undergo a conversion known as metamorphosis, brought about by the heat rising from the Earth's Mantle,

---

*However, as a child, I still remember using calcium carbide in a makeshift lamp, similar to that used in miners' lamps many years ago, adding water to it to produce acetylene. (I also remember the row I used to get because of the soot it created...)

[†]It might be an interesting point to note at this juncture, for those of you who appreciate such facts that, despite the importance of carbon to our survival, it constitutes only 0.008% of the total composition of the Universe. Hydrogen accounts for 92.7%.

turning them into metamorphic* rock. Sometimes rocks are metamorphosed by coming into close proximity to igneous rocks, molten intrusions from the waxy-like substance that lies beneath the Mantle, which, by the way, also drives plate tectonics, another facet of the Earth's geology which has a major bearing on the availability of oil and gas.

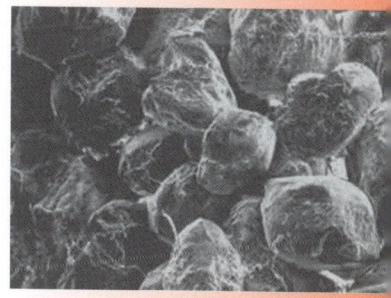

By a process of elimination, we can see that the only rock in which oil or gas can be formed, must be sedimentary. Igneous rocks are out of the question, as they are molten at the time of deposition and, as in the case of basalt or granite, crystalline in nature. Metamorphic rocks may have at one time been the 'kitchen' of hydrocarbons, before they became metamorphosed, but anything included in the rock structure would have been vaporised and destroyed before the rocks reached their final condition.

This micrograph shows what sandstone in a reservoir might look like. Note the void spaces. There is adequate room to hold oil and/or gas. The question is, do these spaces (porosity) connect adequately (permeability) to allow the oil to flow towards the well? One of the inhibitors is what is known as 'wet-ability', whether the individual particles are oil wet, or water wet. If the particles are water wet, the oil will be repelled and will flow more easily out of the reservoir. If the rock faces are oil wet, it will cling onto the oil, making it much more difficult to produce.

Note that I said 'formed in'. Oil and gas can be found in all three rock types, depending on the situation. Here, we digress for a moment and go back to the waxy-like substance beneath the Mantle. Much like the boiling water in a kettle, this molten rock is writhing and rolling, driven by the nuclear furnace that makes up the largest part of the Earth's volume. The movement of this 'plastic' rock is unbelievably slow, but move it does. Piggybacking on it is the Mantle, carrying the continental plates that make up the skin of the Earth and the land on which we walk. The upheaval caused by this movement results in the fracturing and movement of the surface rocks, creating large fissures, uplifting whole sections, dragging others down (earthquakes).

*Metamorphosis is the conversion, through heat, of the chemicals that make up the sedimentary rock. There is sufficient heat to alter the chemistry of the rock, but insufficient to melt the rock completely, resulting in a different kind of rock, e.g., crystalline as opposed to particulate.

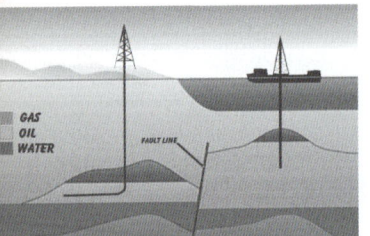

Rock layers are twisted and bent, frequently sheared by tectonic forces, to form traps into which gas, oil and water may accumulate. By knowing where these traps are, mankind was able to drill into them and recover the oil and gas. Pressure, from inflowing water drives the oil or gas towards the upper regions of the reservoir. Gas can be re-injected into the 'gas cap' to drive more oil out. Water may also be injected into the water zone for the same purpose.>>

This picture, of an outcrop in Borneo, vividly shows the power of tectonic forces in being able to literally move mountains. The striated layers are sequential depositions in a water environment, either lake, or sea, of sand and silt, or the skeletal remains of marine creatures which have been compressed, solidified and then upthrust by enormous forces. The rocks shown here are being thrust upwards, they are not slipping downwards into the surrounding rock. The White Cliffs of Dover are an example of upthrust sedimentary rock.>>

Any rock that has open pores, fractures or spaces in it, can become a repository for oil and gas. All that is needed is a seal. That can be another rock, such as shale, which is made of so small particles to be completely water- and gas-proof, a salt plug, which is crystalline and therefore impervious, or another rock which has weathered the upheaval better than the rock below and is not fractured.

The oil and gas seeps mentioned before come from fractures in the rocks overlying the reservoir rock in which the oil and gas has accumulated. Oil and gas seeps still occur today and are used as a means of determining the location of potential hydrocarbon traps.

Hydrocarbons do not exist in pools or caverns as might be though but between the rock particles of sandstones or in the minute fractures that exist in the likes of limestones. What is required of all rock types that hydrocarbons are found in, if they are to be produced, is interconnectivity, or permeability. The porosity of a rock is its inherent space available to a liquid or gas, while its permeability is the amount of interconnectivity that will allow the gas or oil to flow from one part of the rock to another.

When explorers first started drilling for hydrocarbons, the wells they drilled were very shallow, in the order of a few hundred metres. Today, wells are drilled thousands of metres into the Earth, both on land and on sea.

The technology today has been developed to such an art that wells can be drilled many kilometres horizontally away from the drilling rig, increasing the scope and ability of the wells to produce from even remote locations. The driving force behind this development, of course, is cost. Hydrocarbons are still the cheapest form of energy available to Man, even though we know that we are using a finite resource.

As mentioned before, Man's involvement with hydrocarbons stems from millennia before, where their applications ranged from waterproofing of reed baskets, through jewellery making to medication. It was this last use that blossomed in the 17th and 18th centuries. Of course, oils were used for lighting as well, but few attempts were made to refine any of the hydrocarbons dug up.

The Industrial Revolution in Britain and Europe was built on coal as a means of supplying power. Great furnaces belched thick, black, sooty smoke into the atmosphere for centuries, creating the steam that drove the machinery of the revolution. Even today, huge smoke stacks throughout Europe and the rest of the world, can still be seen pouring their noxious fumes into the atmosphere, even though technologies have mitigated much of the more deadly contents of the exhaust streams.

The problem arises from the cheapness of coal, relative to the cost of oil and gas extraction. Most of the extracted oil goes to transportation, not for power. Natural gas availability is still insufficient to replace completely the use of coal as a power source, although slowly, things will change.

As the Industrial Revolution evolved, machines were designed to make other machines. Products, manufactured in mass, had to be moved. Shipping, driven by coal-burning steam engines, began to be developed, but land transport was still highly dependent on the horse drawn wagon.

Going back to the internal combustion engine, a few years after Lenoir's heroic attempts, (I say heroic, because what he did was heroic. If he only knew the risks he took - poor quality steel, questionable finishing, highly explosive mixture - he's lucky he didn't get killed), another Frenchman in 1862, Alphonse Beau de Rochas, a scientist, patented what he called a four stroke engine. Although he never built one, several people after him did, notably Nikolaus August Otto, who built what was to be known as the 'Otto cycle', four stroke engine. This was followed later by the invention by Gottlieb Daimler in 1889, of what became the forerunner of all of today's internal combustion engines.

Another noteworthy was Rudolf Christian Karl Diesel, a German, born in France, who devised the engine that would be named after him, the diesel engine. Basically the same as the Daimler engine, this one used compression to combust the oil used as fuel, instead of the spark plug to ignite the gasoline, as used in the Daimler engine.

So, in a very short period, a matter of 30 years, from Lenoir's engine in 1859, to Daimler's engine in 1889, Man had developed a means of usefully converting hydrocarbons into work. The days of the horse-drawn draught wagon were numbered! Even so, it took a man by the name of Henry Ford and the use of machines and mass production, to get the world off of four legs and onto four wheels. These early vehicles were not particularly efficient, but neither were the steam engines used on the rail system. Although the internal combustion engine was invented and first built in Europe, its use was restricted to gentlemanly races in such gloriously named vehicles as the Peugeot Vis-à-Vis, (four people sat facing each other in a two by two arrangement, similar to the old stagecoaches. Don't ask who drove), Panhard & Levassor, de Dion, at mind-shattering speeds of 12 to 15 miles per hour. Within ten years, as Henry Ford was just beginning to build his Model T, these European engine and car manufacturers were hitting 100 mph, in huge, 14 and 15 litre engines (by comparison, your Nissan or Honda or Ford today probably has a 1.6 to 2.0 litre engine.)

But Henry Ford was more pragmatic. He knew that to succeed in business, he had to mass-produce, at an affordable price. He could hand-build a very expensive car, like Daimler, for the few rich and famous, or he could mass-produce a simple, cheap automobile that even Joe Average on the streets of America could afford. One of his more admirable comments went something like "You can have it in any color you want, so long as it's black". That was Henry Ford.

You could carry coal, like the steam driven locomotives, which cost almost as much to transport as it provided in useable fuel - its power to weight ratio was very low. The internal combustion engine provided a much higher power to weight ratio, i.e., the amount of useful energy derived from a fuel as a function of the amount of fuel that needed to be carried or, in other words, the efficiency of the engine. A steam engine might have an efficiency of 8 to 10%, an internal combustion engine's efficiency was in the twenties.

It is fairly safe to say that Henry Ford played a major part in the development of America as we know it today. Once upon a time, a car was a luxury. It soon became a necessity as the population spread out over the country, opening up the land and allowing the nation to develop into what it is today. That development brought with it an increased demand for energy sources - hydrocarbons especially - as a compact, transportable fuel. You could carry coal, like the steam driven locomotives, which cost almost as much to transport as it provided in useable fuel - its power to weight ratio was very low. The internal combustion engine provided a much higher power to weight ratio, i.e., the amount of useful energy derived from a fuel as a function of the amount of fuel that needed to be carried or, in other words, the efficiency of the engine. A steam engine might have an efficiency of 8 to 10%, an internal combustion engine's efficiency was in the twenties.

Because of this changing form of transport, the demand for hydrocarbons increased tremendously, bringing about a revolution in the oil exploration industry. Already, by 1872 a small strip of land in an otherwise uninhabitable part of Northwestern Pennsylvania, close to where Drake first discovered oil in America, had become the busiest piece of real estate in the US[5]. Townships sprung up everywhere, as companies were set up to exploit this new product, called petroleum, which was bringing millions of dollars into people's pockets. There were no less than three railroads running through the region, railroads that were to become the centre of one of the greatest battles of the 20th century.

Up until the 1850s, the oil being extracted was more of a nuisance than a saleable product. It was being sold quite extensively, but as a medicine or embrocation. Samuel M. Kier, a salt prospector, had already found a use for the oil, '…which came up with the saltwater, was of sufficient quantity to be a nuisance and Kier sought a way to use it. Believing it had curative qualities he began to bottle it. By 1850 he had worked up

A Stanley Steamer - it took half an hour or more to warm up before it would move. Not exactly the car to nip down to the Mall for groceries when you're in a hurry.

Henry Ford with his Model T

The Peugeot Vis-à-Vis

A Classic 'Gentleman's car of the early 20th century

A beautifully maintained 1910 Touring Model T owned by Jacquie & Roland Palmatier, Durham, NH, courtesy of the Model T Collectors' Club

this business until 'Kier's Petroleum', or 'Rock-Oil' was sold all over the United States'[5].*

Then, in 1854, arrived a man who took rock-oil more seriously. His name was George H. Bissell, a one-time graduate of Dartmouth College. On a visit to his old college, he was shown the bottle of rock-oil. His old professor contended that it was as good, if not better, than coal for making illuminating oil.

The oil came from oil springs located in Northwestern Pennsylvania on land belonging to a lumber firm, Brewer, Watson and Company. These springs had long yielded a supply of oil which was regularly collected and sold for medicine, and was used locally by mill-owners for lighting and lubricating purposes.

Bissell, impressed with the commercial possibilities of the oil, at once organised a company, the Pennsylvania Rock-Oil Company, the first in the United States, and immediately leased the lands on which these oil springs were located.

'He then sent a quantity of the oil to Professor Silliman of Yale College, and paid him for analysing it. The professor's report was published and received general attention. From the rock-oil might be made as good an illuminant as any the world knew. It also yielded gas, paraffine, lubricating oil. "In short," declared Professor Silliman, "your company have in their possession a raw material from which, by simple and not expensive process, they may manufacture very valuable products. It is worthy of note that my experiments prove that nearly the whole of the raw product may be manufactured without waste, and this solely by a well directed process which is in practice in one of the most simple of all chemical processes."'[5]

---

* Compare this to the unending legislation still being put in place today, to try and protect people, especially young children, from ingesting hydrocarbon products, as they are considered dangerous and potentially carcinogenic.

Possibly the most prophetic words ever uttered by one person to another. By February 1, 1860, oil was selling at eighteen dollars a barrel! Conversely, by the end of 1861, so many people had rushed to participate in this 'Black Gold' Rush and so much oil was being produced, the price collapsed to ten cents a barrel. So what's new in the world?

A major problem faced by those who were producing the oil was how to get it out of one of the most inhospitable places in the Americas. The obvious answer was to put it into barrels* (hence the term 'a barrel of oil') and take it out by horse and wagon. This turned out to be a very lucrative venture for the farms in the area, as the farm boys hitched up their teams and offered their services. Sometimes, as many as a hundred wagons could be seen on the road at a time, all hauling barrels of oil. These teamsters were not averse to a little brawling once in a while to get what they wanted, and were notorious for their 'black snake' or long, leather horse whip, which could soon change a person's mind.

But events were to catch up on the transportation of this new product. The first pipeline to be built and operated infuriated the teamsters who immediately tore it apart. The next pipeline built, they actually set on fire, along with the tank farm, which was used for holding the oil prior to it entering the pipeline. Eventually though, the pipelines won out and the teamsters had to give up, ungraciously in the most part. They were losing out to a much greater force, the railroads.

Although oil was first discovered in Pennsylvania and the first blossoming of the industry began there, because of the better connections and transportation systems from another state to points beyond, it was not long until a large refining industry evolved in Cleveland, Ohio. Its thirty something refineries were selling products all across the western seaboard while Pennsylvania was still struggling with moving its oil from wellhead to market.

---

*A barrel of oil is, by definition, 42 US gallons. There are 1.200952381 US gallons in an Imperial gallon.

One young Cleveland man who had an eye on the oil-refining business, had remarkable commercial vision - a genius for seeing the possibilities in material things. This man's name was John D. Rockefeller. Although he was only twenty-three years old when he first went into the oil business, he had already got his feet firmly on the business ladder, by his own efforts.

Early on, he learned the value of investing money: "Among the early experiences that were helpful to me that I recollect with pleasure was one in working a few days for a neighbour in digging potatoes - a very enterprising, thrifty farmer, who could dig a great many potatoes. I was a boy of perhaps thirteen or fourteen years of age and it kept me very busy from morning until night. It was a ten-hour day. And as I was saving these little sums I soon learned that I could get as much interest for fifty dollars loaned at seven per cent. - the legal rate in the state of New York at that time for a year - as I could earn by digging potatoes for 100 days. The impression was gaining ground with me that it was a good thing to let the money be my slave and not make myself a slave to money."[5]

By 1870, through astute business sense and not a little brass neck, John D. Rockefeller combined the two oil refining companies and one marketing company in which he had either a controlling interest or he owned, into one - The Standard Oil Company.

"**This is a story** of a **human enterprise** that has shaped and will continue to shape civilisation... Those who got rich became filthy rich..."

Michael Economides & Ronald Oligney in 'The Color of Oil'[6].

02

THE MAKING OF
THE MAJORS

Rockefeller's Standard Oil Company started business with US$ 1 million capital, a large sum of money even today. Rockefeller made much of his money by squeezing rebates out of the railroad companies to whom he gave his business of transporting oil and products. Before 1870 however, others were figuring out how this man could undercut them in the open markets, because everyone bought their oil at much the same rate, and started squeezing the railroad companies as well. Soon, the railroad companies would be at loggerheads over shipping prices, in an effort to retain business.

In the midst of it stood John D. Rockefeller, the biggest producer from the Cleveland area, controlling the discounted shipping cost he was getting from Lake Shore Railroad, which he had brought into the oil transportation business in the beginning.

"Not only did Mr. Rockefeller control the largest firm in this most prosperous centre of a prosperous business, he controlled one of amazing efficiency."[5]

In the creation of the Standard Oil Company, Rockefeller had brought together a group of remarkable men; Samuel Andrews, by all accounts the ablest mechanical superintendent in Cleveland; William Rockefeller, his brother, not only an energetic and intelligent business man, but a man whom people liked; and probably the strongest man in the firm after John D. Rockefeller, Henry M. Flagler. According to Ida Tarbell's description, brother William was open-hearted, jolly, a good story-teller, a man who knew and liked a good horse - not too pious, as some of John's business associates thought him, not a man to suspect or fear, as many a man did John. Oil men would tell of how much they liked him in the days when he used to come to Oil City buying oil for the Cleveland firm. The personal quality of William Rockefeller was a strong asset of the Standard Oil Company. It must have been evident to every businessman who came in contact with the young Standard Oil Company that it would go far.

The company, itself, should have known it would go far, but John D. Rockefeller was far from satisfied. He was a brooding, cautious, secretive man. He saw all the possible dangers as well as all the possible opportunities in things. He studied, as a player at chess, all the possible combinations, which might imperil his supremacy. His biggest concern was how to maintain his lead against the Cleveland refiners, as well as the refiners near the wellheads, with a refining capacity now equalling that of all Cleveland. Boasting that they would soon be shipping oil to the world, the creek refiners were being encouraged by the Pennsylvania Railroad. Could Standard Oil stand against such competition? He was not alone in this concern either. The Lake Shore and New York Central Railroads stood to lose a profitable business sector to a rival railroad, Pennsylvania.

Several things preyed on the mind of young Rockefeller. With the increase in refining, production was now wildly in excess of demand (remember, we are still talking about medication, lubrication and lighting products. Gasoline wasn't even a dream yet), with the result that prices were falling. Production costs had fallen, to be sure, but so too did profits. Rockefeller was a profits oriented businessman.

He also was worried, as were the railroads of Philadelphia and Pittsburg that much of the export was now as crude. Overseas buyers were quick to fathom that it was cheaper to buy the crude and refine their own products. The refineries we are talking about in those days were little more than a glorified 'white lightnin'' or 'moonshine' still, with a copper kettle, fire and cooling tubes, not the multi-billion dollar castles of the oil industry that we see today. Some countries, France for instance, were even imposing import tax on refined products to protect their fledgling refining industries! This was the 1870s!

The solution, when first mentioned to Rockefeller, did not appeal, as it also did not appeal to his partner Flagler. But the pressure from all sides, including the railroads and several other refiners convinced him that it was indeed the right move - the formation of a consortium of refiners and transporters, that would be so big, command such a large slice of the market, it could control transportation rates, control product production (and hence price) - in essence, control the market. More importantly, controlling the market, it could force the railroads not to carry crude for export, so that only refined products could be exported.

"It was evident that a scheme which aimed at concentrating in the hands of one company the business now operated by scores, and which proposed to effect this consolidation through a practice of the railroads which was contrary to the spirit of their charters, although freely indulged in, must be worked with fine discretion if it ever were to be effective."[5]

The first thing that Rockefeller did was to buy up the charter of the Southern Improvement Company, also referred to as the South Improvement Company. This gave him the right to 'carry on any kind of business in any country and in any way'. Secrecy assured, he then required of all parties to the deal a declaration of their loyalty and sworn secrecy to the company.

The railroads balked at any idea that could put anyone out of business, as it would in the end affect their business, and so, over the next few weeks almost every refining company was coerced into joining the South Improvement Company. Divided into 2000 share lots, the largest shareholder was none other then the Standard Oil Company, with 900 shares, held between the stockholders of Standard. In actual fact, although they managed to convince the railroads that they controlled the refining industry in America, they only controlled something like 10 percent of the industry.

But the system worked. Soon, the South Improvement Company had all the railroads tied into their agreement and were given specialised transport rates. Not only that, they got a cash refund on any oil shipped by refineries not in the consortium, as they were charged a higher rate. Even more useful, the consortium got daily reports from the railroads of who was shipping what, where, giving them invaluable insight into the activities of their competitors! And their leverage over the railroad companies was the amount of business they were able to offer them in years to come. A change of mind by the South Improvement Company could be the difference between life and death to a railroad company.

Rockefeller had by now increased both the shareholders in Standard Oil and the paid up capital. He then proceeded to visit the individual refiners in the consortium offering them shares in Standard Oil if they signed over their rights to Standard Oil. Since these were the very refiners Rockefeller had squeezed with his preferential shipping rates from the railroads in the beginning, many of them balked at the idea but eventually better than 80 percent of them signed up.

"Under the combined threat and persuasion of the Standard, armed with the South Improvement Company scheme, almost the entire independent oil interest of Cleveland collapsed in three months' time. Of the twenty-six refineries, at least twenty-one sold out. From a capacity of probably not over 1,500 barrels of crude a day, the Standard Oil Company rose in three months' time to one of 10,000 barrels. By this manoeuvre it became master of over one-fifth of the refining capacity of the United States. Its next individual competitor was Sone and Fleming, of New York, whose capacity was 1,700 barrels. The Standard had a greater capacity than the entire Oil Creek Regions, greater than the combined New York refiners. The transaction by which it acquired this power was so stealthy that not even the best informed newspaper men of Cleveland knew what went on. It had all been accomplished in accordance with one of Mr. Rockefeller's chief business principles - Silence is golden."[5]

But Rockefeller was not to have things all his own way all of the time. The South Improvement Company was still working away at the refiners outside of Cleveland, trying to persuade them to join the consortium when a series of human errors let slip details of the proposed freight rates that were to be applied to those that did not sign up. Within twenty-four hours an angry mob was knocking on the door of the Legislature, demanding that the South Improvement Company's charter be withdrawn while another group lobbied Congress to have the whole scheme investigated.

There then began the first of what was referred to as the 'Oil Wars of the 1870s' - refiners set to discredit and disband the South Improvement Company. Rockefeller was made of much sterner stuff than the others ever imagined however and using every business tactic in the book he eventually prevailed, setting himself up in a position where Standard Oil was more or less in control of all the refining industry in the country. His approach was simple, still very much the same theme as used by the South Improvement Company - join us, or we'll put you out of business.[7]

The motivation for Rockefeller's actions was, in the beginning at least, probably honourable. He was after all, an astute businessman and he saw in the merging of all refining and transportation of crude oil in the States into one large organisation, the creation of an entity that could control not just the American industry but also the global industry. And that, basically, has been the history of the oil industry ever since, the need for control, because of the vast sums of money to be made in the business. Rockefeller simply was the first. He beat OPEC (the Organisation of Petroleum Exporting Countries) by more than sixty years in that respect.

What followed gave rise to words like 'Trust Busters' as the American people, the American Government, took Rockefeller to task and tried to break up his empire. Rockefeller was unquestionably the first American billionaire and the family, still to this day, is heavily involved in American society and commerce. Rockefeller just happened to be the right person, in the right place at the right time. He controlled the majority of the refining in the States just at the time that Henry Ford began rolling his model T Fords off the end of the conveyor belt. They needed fuel, he could supply it.

By 1911, when the Trust Busters finally succeeded in tearing apart Rockefeller's empire, others were already forging their own empires in the oil industry. As early as 1901, a wealthy investor, already rich from gold investments in Australia, by name of William D'Arcy had signed a concession from the then ruler of Persia, His Majesty Muzaffar al-Din Shah[8] to explore and exploit all of Persia, except the five most northern provinces, in his search for oil.

The motivation of both gentlemen was pure and simple - money - the Shah needed money to keep his Persian Empire together, while D'Arcy, already very wealthy, wanted more.

Although things started off slowly, with many impediments in his way, including trying to form agreements with the local tribesmen, by 1903, D'Arcy was convinced he could do business in Persia and swiftly formed a company called First Exploitation Company.

However, as is the way of the oil industry, not everything goes smoothly, especially in those days, and before long he was being pressured by Lloyd's Bank to settle his overdraft, which stood at £150,000 or about £25 million in today's currency. Desperate to keep his enterprise going, D'Arcy eventually approached Burmah Oil, a very successful British oil company, (second largest only to Shell), that had made good in the oil fields of Burma, for financial backing.

Even Burmah Oil found itself financially stretched because of the problems of drilling in the inhospitable deserts of Persia and had ventured its final offering of £40,000 towards helping D'Arcy making his dream come true, when the Majid-I-Suleiman well blew in. Now convinced that there was oil to be found in the region, Burmah Oil immediately began "a prospectus rapidly being written for a new company to raise capital for the development of the oilfield".[8]

The new company was the Anglo-Persian Oil Company.

Although the launch of the new company raised more than enough capital to pay everyone off handsomely for their risks, it did not solve all the problems, one of which, was the rapidly accumulating stockpile of oil, which could not be got rid off. The company had battled long and hard on the logistics of trying to move oil in volume over such inhospitable terrain, to be solved eventually by the installation of a pipeline to Abadan at the head of the Persian Gulf. Highly sulphurous, the oil was also a problem in that it was difficult to refine, and therefore sell. In desperation, Anglo-Persian had to sell off excess oil at a price of two shillings* (10p today) a barrel, much to the chagrin of all concerned except the board of Asiatic Petroleum, a Royal Dutch-Shell subsidiary, to whom they sold it. Asiatic Petroleum was Burmah Oil's main Asian competitor.

Money was running out, and Anglo-Persian had no choice, as they feared they might not be able to stop a bid by Shell to buy into the potentially lucrative Persia industry. The very first shipment of oil from the Anglo-Persian oil fields, 15,000 barrels, left Abadan on a ship under charter to Royal Dutch-Shell. Such were the vagaries of war and business.

Who knows what might have happened if, at that moment in time, the British Royal Navy had not stepped in on Anglo-Persian's behalf? Under pressure from Winston Churchill, then First Lord of the Admiralty and a firm believer in the use of oil as a fuel for the new Navy being built, the Government of Great Britain signed two agreements with Anglo-Persian. The first gave the company a much-needed cash injection and, more to the point, gave the Government a controlling interest in the company. What the British Government wanted - needed - was to retain a firm hold on Mesopotamia and a ready source of fuel for its Navy.

Even so, even at that time, news reporters were warning that, "…the oil wells of the Royal Navy will be an abiding temptation in times of trouble…"[8] Lacking parliamentary approval of the accord, pressure was put on the Government to do so immediately.

On the basis of Standard Oil's hold over the western hemisphere and Royal Dutch-Shell's grip on the Old World, Churchill asked the Government to pass a bill on the acquisition of Anglo-Persian Oil Company, because

---

*To put this in perspective, there was no OPEC to set prices in those days and it was a sellers', or a buyers' market, depending on how desperate the other party was. Rockefeller sold oil at US$ 20 a barrel in the 1860s! He also sold it at a few cents a barrel by the 1870s!

"...the oil consumer has not got freedom of choice in regard to other alternative fuels, but neither has he freedom of choice in regard to the sources of supply from which he can purchase." The sole independent voice was that of Burmah Oil and its offshoot, Anglo-Persian Oil Company. The resolution was passed by 191 votes to 67.

Not only - by diluting its holdings - did this relieve the pressure on Burmah Oil which, as a major shareholder in Anglo-Persian Oil, was in imminent danger of being forcefully taken over by Royal Dutch-Shell, it provided desperately needed finance to the company. The amazing thing about it all was that management control remained with Burmah Oil, despite it now being a minority shareholder!

On the 28th of June, eleven days after Churchill pushed the bill through parliament, the Archduke Franz Ferdinand, heir apparent to Emperor Franz Josef, was assassinated in Sarajevo, leading to Germany's full-scale attack on France six weeks later. The Great War had begun.

It is a well-known fact that in war, there are vast profits to be made by some. Anglo-Persian was one such company that benefited greatly. At the beginning of the Great War, the company was producing around 5,600 barrels of oil a day. By the end of 1918, it was flowing 18,000 barrels a day into its pipelines. Not without a lot of blood, sweat and tears, from all concerned, it must be added.

Charles Greenway, who had become chairman of Anglo-Persian, felt that, as sea freight costs were going through the roof, the company would be better off investing in its own tankers. By 1918, the British Tanker Company, the new subsidiary of Anglo-Persian, was very successful. Greenway saw an opportunity to get his hands on a British sales and distribution company, a subsidiary of a German company, Europäische Petroleum Union. It held the agency to distribute Shell products in the UK and, by 1914 held 36 percent of the British petrol market. Greenway bought the company, which because of its German beginnings had been classified as an enemy concern, from the Public Trustee for Enemy Property. The company's name was British Petroleum.

With the acquisition of British Petroleum, Anglo-Persian moved its majority assets out of Persia and into downstream processing and transportation. Still, with only one oilfield - Majid-I-Suleiman - this did not make it the major oil company that it is today. By the end of the Great War there were some 48 wells producing around 19,000 barrels a day. Duncan Garrow, the company's technical director, not convinced that the reservoir was all one huge structure, wanted more. Much more. He anticipated 76,000 barrels a day after a crash drilling programme of 123 wells over the next few years, to ensure the company's continuity. Luckily, sanity, in the form of oilfield manager R. R. Thompson prevailed, arguing that drilling more wells would simply deplete the structure faster. Eventually the Majid-I-Suleiman reservoir was proved to be a single structure, by a new technical advisor, one John Cadman (later to take over the company chairmanship from Greenway).

To augment the company's oil resources, Cadman called in an eminent geologist from Hungary, Professor Hugo de Böckh, who advised the company to drill in certain areas of the country, three of which Haft Kel, Agha Jari* and Gach Saran, proved to be the equals of Majid-I-Suleiman. Anglo-Persian/BP had found its assuring oil resources.[8]

Another of Professor Hugo de Böckh's 'discoveries' was in Iraq, where Anglo-Persian had a vested interest (47.5 per cent to be exact) in the Turkish Petroleum Company. An Anglo-Persian director, Herbert Nichols became managing director and acting director of the petroleum company and immediately began exploration activities in Iraq. Within six months, at Baba Gurgur, just north of Kirkuk, an immense discovery was made, turning Iraq into the most valuable asset in the world. Although not the first - Anglo-Persian had made previous discoveries in Iraq - it was unquestionably the biggest. Anglo-Persian, or BP, had arrived.

*Having worked on a blowout (a 300 ft high gas flame) in the Agha Jari field, I can attest to the size of the field!

02

Together, with Exxon (or Esso - the original Standard Oil Company of New Jersey), Shell (Royal Dutch/Shell), Gulf, Texaco, Mobil (Socony - or the Standard Oil Company of New York) and Chevron (or Socal - Standard Oil of California), BP went on to form what became known as the Seven Sisters.[9]

Then, even as today, assets were everything, and the Seven successfully carved up the world's producing areas amongst themselves, at least in the non-communist countries. But their very existence caused many of the oil-producing countries to rebel against their domineering attitude towards production and pricing. In 1975 these countries called the first Sovereigns and Heads of State Summit of a little know (outside of the industry at that time), organisation - OPEC, or the Organisation of Petroleum Exporting Countries.

OPEC's origins lie, in 1949, in an attempt to establish more useful communications between some of the oil producing countries, Venezuela, Iran, Iraq, Kuwait and Saudi Arabia. As oil companies continued to change oil prices to suit their own needs, these five countries held the first true conference of the organisation in September 1960, establishing OPEC as an intergovernmental agency.

By 1975, OPEC had grown to include Qatar, Indonesia, Libya, the UAE, Algeria and Nigeria. Two other countries - Ecuador and Gabon - although originally made full members in 1972 and 1975 respectively, both later terminated membership at their own request. At the moment of writing, it appears likely that Indonesia will also terminate membership sometime in the future.

What had, on occasions, been a political tool in the past, most definitely became a political tool on the formation of OPEC. The member countries had every right to try and gain control of the precious liquid that was being extracted from under their feet. However, according to the Organisation:

> 'Throughout its existence, *OPEC* has promoted the ideals of the UN, a landmark being the 'Solemn Declaration' adopted in

1975 by the First Conference of Sovereigns and Heads of State of *OPEC* Member Countries, in which they stress that world peace and progress depended on mutual respect for the sovereignty and equality of all members of the international community, in accordance with the UN Charter, and emphasised that the basic statements of the Solemn Declaration fell within the context of the decisions taken at the Sixth Special Session of the UN General Assembly on problems of raw material and development.'[10]

OPEC countries today answer about 40% of the world's demand, a figure likely to increase to 50% before the end of the decade (2010). With some 75% or more of the know oil reserves left in the world, OPEC is in a position to dictate terms. This, in turn, leads us right back to the state of affairs that existed at the time the Seven Sisters carved up the oil world at the beginning of the 20th century. Nothing changes. Mankind has been fighting over this black gold since time immemorial - why should we expect it to change now?

" **In this era of global change** we face what may be humanity's greatest crisis… for over 99% of the history of humanity we lived as low-density foragers or farmers in egalitarian communities of no more than a few dozen persons… such a cultural system, based primarily on human labour, can generate only about 1/20* horsepower per capita per year…" [11]

*Compare this to the hundreds of horsepower generated by some road cars today or the thousands of horsepower needed to fly the new A380 Airbus.

03

But this is a book about energy. Not about politics, not about prices or control, but about energy. However, the preceding chapter serves to show how sensitive an issue energy can become. Without energy, the world - Mankind - would not have developed to the extend it has. Going back to a previous footnote, if you remember the figures, around 10,000 years ago the global population was something like five million. The majority of capital cities in the world today contain more people than there were in the whole world in those days! Look at Tokyo, peaked at 12 million people, New York eight million and counting.

Even by the middle of the 17th century, the world population was only about 500 million souls. That's a hundredfold increase in around 9600 years. Given the prolific nature of our species and the obvious improvements in things like food availability, medication, however crude, this is not really surprising. What really is surprising is that in the following 400 years the population jumped to six billion, a 1,200-fold increase. Talk about exponential growth!

Now something serious had to have caused such a sudden increase – the human population, like a viral infection in an unprotected body, suddenly exploding. What caused it?

Energy, or more specifically, the Hydrocarbon Era. The last hundred years or so has seen the most rapid increase in population of any mammalian species on this planet. And the cause of it can be laid fairly and squarely at the feet of the Hydrocarbon God that we all, even to this day, worship.

Don't believe it? Take a look around. If you are in an industrialised country, you are benefiting from all the trappings that energy availability can bring you. If you are in a third world country however, maybe even an oil producing country, you may be lacking many of the simplest benefits that energy can provide you, such as fresh water, electricity, sanitation, things that most of us take for granted but which our ancestors would have

One of the curses of hydrocarbon energy is that it is so easily produced, it is very portable and it is highly adaptable to the many needs of Mankind. Despite the enormous quantities of oil being consumed, approximately 85 million barrels a day, the majority of production goes into transportation - petrol (or gasoline) engines, diesel engines, kerosene for aviation fuel, etc., and we've already seen how inefficient these engines can be.

been hard pressed to produce from dried cow or elephant dung, brush or drift wood. Even today, almost 1 billion people in the world suffer from malnourishment with limited or no access to food or even water while almost half the global population survives on less than US$2.00 a day. These people want, and deserve to share in the benefits that hydrocarbons have bestowed on Mankind.

Hopefully the above has given some insight into how we got 'Here' - the present situation we face in the world today. Which leads into the next topic of this book:

■ What we are doing here

One of the curses of hydrocarbon energy is that it is so easily produced, it is very portable and it is highly adaptable to the many needs of Mankind. Despite the enormous quantities of oil being consumed, approximately 85 million barrels a day, the majority of production goes into transportation - petrol (or gasoline) engines, diesel engines, kerosene for aviation fuel, etc., and we've already seen how inefficient these engines can be.

But when you have an inexhaustible source of energy, what is the point of worrying about efficiency? Empires have been built, literally, on the belief that oil was inexhaustible. Companies like Exxon, Shell, etc., have annual revenues greater than some countries. With that money, they encouraged the use of oil, in its many shapes and forms. The global economy depends on oil for its continued wellbeing. But that source is very definitely not inexhaustible, as more and more the experts are trying to tell us.

Although the vast majority of the hydrocarbons produced do go into transport, industry has developed many other products from the self-same hydrocarbons. Your plastic covered polystyrene tray of food sealed with polyethylene, to be found on many a supermarket shelf, owes its existence to hydrocarbons. The crumple zone on your car (the plastic fenders etc.,) comes from hydrocarbons. Even the lipstick that women use owes its quality to hydrocarbons.

Its very versatility makes hydrocarbon a product very much in demand. From previous chapters, you learned that both carbon and hydrogen are very amiable when it comes to sharing electrons with other atoms and compounds.

Of the many millions of compounds known to Man today, the vast majority of them are related to the hydrocarbon chain. Where should we start listing them? Should we start with foodstuffs or fertilisers? Acids or alcohols? Plastics or polymers? Wherever you turn, the chances are that you are directly involved with hydrocarbons in some way.

Sitting here at my desk typing this manuscript, I can think of the keys on which I press – some form of plastic. The chair I sit on, a plastic base and a hydrocarbon-derived cloth to cover it. The foam inside is derived from hydrocarbon. The electricity that enables my computer to work is converted from natural gas, part of the hydrocarbon family.

My TV, my couch cover, my colour printer, the fridge, the food containers within the fridge, the washing machine and a lot of the clothes within it, the bottle containing the detergent and the detergent itself – they are all made of hydrocarbon-based materials. Without hydrocarbons, all I would have is four bare walls and a heap of rubble in the middle of the floor. I am very much dependent on hydrocarbons for my lifestyle. And so is the majority of people in today's modern society. Look at the table in Appendix I and see the relationship between hydrocarbons and our everyday products.

We have followed willingly, the path towards a throwaway society. Originating, as many such things do, in the United States – the home of instant gratification – disposable goods save a lot of personal inconvenience. Drink cartons, plastic bottles, supermarket carrier bags, food trays, drink cups, you name it. If it provides a convenient solution to a daily problem, it will be embraced with open arms. No dishes to wash, no containers to carry home, just put it in a plastic bag and trash it. End of story.

Well, not quite. The energy used to produce those throwaway items, and the items themselves, used up a lot of the hydrocarbons that could have been put to better use. Like providing electricity to some destitute villages in Africa, or Ethiopia, or Afghanistan, or wherever that are desperately in need of the basics of sanitary living, like running water.

One of the things that hydrocarbon energy has done to our lives is widen the gap between the 'haves' and the 'have nots'. We, the rich world, have an insatiable appetite for hydrocarbons, for which we are willing to go to war, to devastate large tracts of other people's lands, to ensure our constant supply. We will still have our fast cars, our transcontinental flights, our throwaway society. Of course, we will pay these countries for their products, these countries that we have decimated in the process of getting the hydrocarbons, so that they too will be able to build a society based on ours – once they have paid for the rebuilding of their country and paid off massive world debts incurred while trying to protect their country in the first place.

We all would like to live the Great American Dream – nice house, swimming pool, big car, 42 inch TV, etc. But that is exactly what it is – a dream. Only a minority of Americans actually live the dream, the dream that is spoon fed to us through TV programmes, films and magazines. But how long can we continue to live this dream? Let us go back to basics and look at society and how it developed and see why the dream cannot go on indefinitely.

Man is primarily a hunter-gatherer. Our ancestors hunted (and were hunted in return) for food. We gathered berries, grasses, etc., to form a staple diet, enough to keep a few dozen people alive. We roamed the plains in small groups – remember, five million in the whole world 10,000 years ago? Half a million years ago as our ancestors were just climbing to the top of the evolutionary tree, how much fewer must have been the numbers of the hunter-gatherers? The men hunted and defended, the women gathered and also looked after the children. It was a tough existence and life expectancy was probably mid-30s for a healthy individual.

Slowly however, agriculture and domestication of animals took over as the preferred way of living, creating villages, small communities of like-minded people. These groups needed protection, both from other groups as well as from wild creatures while they tended to their crops. Defending the state was a full time job, requiring specialised people. Mostly young warriors, these defenders expected support in the form of clothing, food and other basic necessities from those they were protecting. A barter process began to evolve, 'I'll protect you if you feed me' type of arrangement.

One of the things that living in a protected environment did for Man was it allowed him to multiply more successfully. As a hunter-gatherer it was difficult to look after more than one or two children at a time. Wild animals preyed on any young left unattended. Living in an agricultural commune, with its increased protection, there was also likely to be more food available, children survived longer. Children were essential to the community because, as the community grew, more energy was needed, to plant, to tend, to herd - and the only energy available to our ancestors was the labour of their own backs. Another thing that the success of the commune did was to free some people from their daily labours to allow them to practice an art or craft that they excelled in – pottery, weaving, etc., which they could then barter with others to support themselves. One person could provide enough food for two, freeing the second person to increase the common wealth.

As the village economy grew, so did the production of items for barter, leading to a trading system with other like-minded villages, a trading practice that extended further and further afield, creating a basic global economy, not too dissimilar to the one we practice today. The advantages were obvious. By trading with others, the spread of innovation improved the lot of not just those who created it, but everyone who could afford it. The beginnings of mass production were in the pipeline.

The villages proved to be very attractive to others as well. Ancestral hunter-gatherers who were still struggling against nature found that moving into

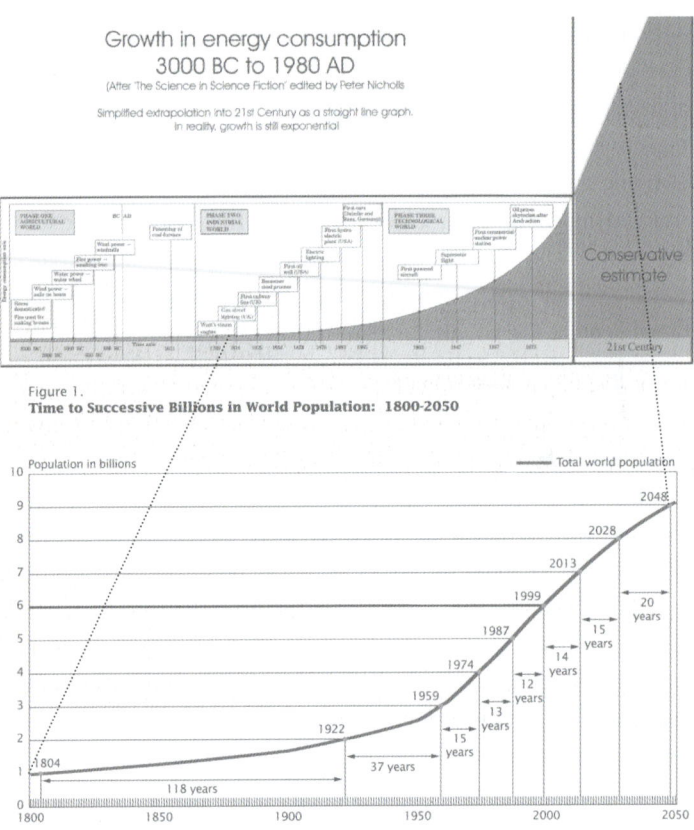

Figure 1.
Time to Successive Billions in World Population: 1800-2050

Source: United Nations (1995b); U.S. Census Bureau, International Programs Center, International Data Base and unpublished tables.

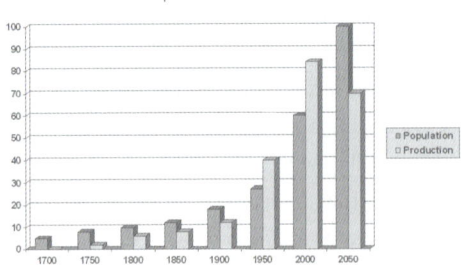

Although on different scales, the similarity in trends (see dotted lines) in the upper and central graph are unmistakable. Both increase rapidly from around 1800. The lower graph shows that by 2050 if not before, our population will have completely outstripped oil production

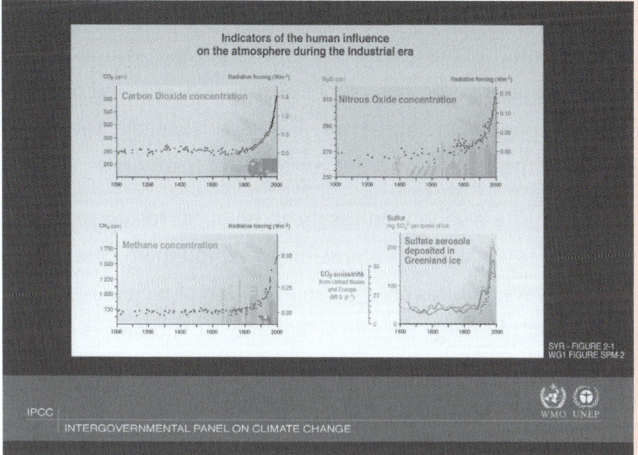

This image shows the relationship between human population growth and atmospheric gases build-up. Compare this to the graphs on the previous page.

the villages provided them with security, food, etc., provided they had something to offer the village in return. As our ancestors became more and more an agrarian society, so the demand for energy increased – to plough, to grind corn, the strength of the ox and horse were brought into use.

And so it went on. Each incremental improvement allowed the freeing of human labour from the basic necessities to be applied elsewhere. Each new discovery - such as iron - required more energy. Before iron could be utilised properly, fire had to be controlled. It is thought that our predecessor, Homo Erectus, controlled fire as much as 400,000 years ago. The controlled use of fire led to many useful products, not least of which was the fire brick, albeit hundreds of thousands of years later! A house built from fire bricks lasted much longer and afforded the owner considerably more protection than one built from mud, logs and other detritus. This led also to the creation of more permanent establishments.

But in all this time, the basic source of energy utilised by Man was the labour of his own back, animals or some form of energy that he could produce from other sources, such as fire. Which is why, even until relatively recently, (in the history of Mankind at least, assuming his, or his predecessors presence on this planet of between one and a half to two million years) population growth rate has never exceed a few percent per annum – until the discovery of the versatility of fossil fuels.

'Since nothing is a resource unless it can be used, resources are defined by the technology that makes it possible to exploit them. Since exploiting a resource always requires energy, the evolution of technology has meant the application of energy to a growing array of substances so that they can be "used to advantage". In the brief time since humans began living in cities, they have used more and more energy to exploit more and more resources.' [1]

The availability of this versatile energy source is behind the massive explosion of population. Today, we number some 6.6 billion people (for an accurate estimate, check on *http://www.census.gov/cgi-bin/ipc/ popclockw* through your Web browser). We live in skyscraper buildings, we drive powered vehicles on hard-topped roads, we eat foods from almost anywhere in the world, we commute from one part of the globe to another with as much ease as our grandfathers commuted from one regional city to another, something their grandfathers would have considered as a major event in their life. Ten thousand years ago, it would have taken our ancestors, if it had ever crossed their minds, several generations to move from Western Europe, to Eastern Asia. Today, we do it in a matter of hours.

We all belong to one huge global city now. Communication has provided us with a means towards instant gratification – you can see something developed in Japan on your TV screen in London or Chicago and demand it from your store before it has even hit the shelves in Tokyo. The media has brought material benefits to our doorsteps. We have become a consumer society, a materialistic society – a must-have society.

Look at your newspaper. Look at your television programmes – tempting you, taunting you, a new camera, a new handphone, new furniture, new clothes, new car, overseas holidays – you name it, it's being touted by your media as something you simply MUST HAVE. "This programme is being brought to you by XYZ Toasties – the ONLY way to eat breakfast!! Get some today before you forget!!"

How can you possibly forget when it's being rammed down your throat ten times an hour and appears on every third page of your favourite magazine?

And how can we manage to create all these products, which in most cases we really don't need? Because of the availability of energy. We are squandering our birthright. The amount of oil consumed in one month today would have lasted the best part of a year in 1950 and yet some 40% of the world population still doesn't have access to electricity.

Going back to the quote from Dr David Price's report at the beginning of this book, "…indeed, the ability to use energy extrasomatically enables human beings to use far more energy than any other heterotroph that has ever evolved…" it was this ability to find uses for energy far beyond our basic survival needs that set us apart from all other forms of life in the past, now and in the future. The problem is, if governments around the world stick with current policies the world's energy needs will be well over 50% higher in 2030 than today. More about that as the story unfolds.

"**People were terrified.** They sped home to loved ones or searched for a priest. The mass hysteria even caused some individuals to tell reporters and police that they 'saw' the invasion."

result of Orson Welles' radio presentation of H. G. Wells' 'War of the Worlds'.

04

WHO IS TO BLAME?

Media, especially television and the Internet, has allowed us instant access to even the remotest parts of the world from our armchairs. George Bush Snr, as President of the United States of America, is reported to have heard of the launch of Operation Desert Storm (the Gulf War) before even his generals had informed him, from CNN News Channel on his TV. And this can and has, set dangerous precedents, allowing uninformed people to make far-reaching decisions based on minimal information - information which is frequently incorrect.

Few people, outside of the immediate group involved in the exploration for, the production and the processing of hydrocarbons, know or care very much about it. Hydrocarbon production is today a very highly developed technology and one of the things that sets technology apart from all other aspects of human endeavour is the factual reality of that technology – it is based on fact. It is based on proven experiment, tried and tested means of measurement, it is based on Fact. There is no room for speculation or hearsay - there can and will be differing opinions, but the opinions must be based on known facts and are simply a difference of interpretation of these facts. Anything other than that exposes alternative ideas to be the basis of a lack of knowledge. And this is the source of incorrect information.

Let us sidestep for a moment, and go back to the time of the formation of these hydrocarbons. Scientists, botanists, geologists etc., are pretty much in agreement as to how coal and hydrocarbons were formed. Although there are still some people who maintain that hydrocarbons are spontaneously created within the bowels of the Earth*, most scientists

---

*This is similar to William Thompson, Lord Kelvin's refusal to accept the age of the Earth as being anything other than the 24 million years he had calculated, even after Rutherford and others had given irrefutable proof, through the discovery of radioactivity that the Earth had to be at least 700 million years old [the half life of uranium (found in pitchblende), the only radioactive element on Earth with a half life long enough to be still around in measurable quantities]. Subsequently, with the identification of Cepheids, Edwin Hubble removed any doubt as to the age of the Universe – it was billions of years old.

agree as to their formation. It took some 400 years of 'primary production', i.e., plant growth through photosynthesis, death and anaerobic burial, for the amount of fossil fuels burned in 1997 to be produced.* The gallon of gasoline you burned in your car today required approximately 90 metric tons of ancient plant matter as precursor material to make it. It therefore doesn't require a mathematical genius to figure out that in 100 years we have consumed many hundreds of thousands of years of photo-synthetically captured solar energy. This year's consumption of energy will require more than 400 years of growth to replace, at standard primary production rates!

But not only are we taking with the one hand, we are destroying with the other. During the Carboniferous period there were no extrasomatically active creatures around to make use of the fallen trees, ferns etc., so they had time to accumulate, consolidate and become buried. Today, the human race is so numerous that not a leaf is left unturned. We and our ancestors have burned everything we could get our hands on. We are felling huge forest giants to make garden furniture; we are clear-felling enormous swathes of land for agricultural use to feed the burgeoning masses. There is no replacement of the carbonaceous material that was buried so long ago and which we are now using at an increasing, and increasingly alarming, rate.

Although absolutely accurate figures take some time to appear in the oil and gas industry (sometimes they never appear, for the simply reason that knowledge is power and power is money…) we are consuming something in the region of 86 million barrels of crude oil a day on a global scale (2007). New, replacement discoveries run around 25 to 35 million barrels a day. To remain where we are today, with no further increase

---

*Jeffrey S. Dukes, Department of Biology, University of Utah; 'Burning Buried Sunshine: Human Consumption of Ancient Solar Energy', Kluwer Academic Publishers

in consumption, we would need to be discovering 86 million barrels of crude oil a day! I'm not talking about producing from existing reserves, I'm talking about finding NEW sources every day.

Shell discovered the Malampaya field offshore the Philippines, a huge gas reserve with enough gas to provide for power generation on the island of Luzon for twenty years or more (provided the Philippine population remains around the level it is now and no increase in demand is made). Beneath the gas in the reservoir, Shell also discovered an oil rim thought to contain some 200 million barrels of oil, a not insubstantial volume where the Philippines is concerned. But the pressure to drive out the oil came from the gas that was already being produced so, at the government's request, the company fast-tracked a development well and put the oil rim onto what is called an Extended Well Test (EWT), where the reservoir is tested while producing oil at the same time. After producing several hundred thousand barrels the EWT was halted and no further decisions were made about its future at that time. One of the reasons Shell was hesitant to spend large sums of money in production of the oil is that even if the reservoir contained an estimated 200 MMb (million barrels) of oil, they would be lucky to produce (under natural drive conditions alone), more than about 25 to 30 percent of that oil, or about 60,000,000 barrels, about a hundred times what they produced in the EWT – but more importantly, less than one day's global consumption!

And therein lies a clue to one of the biggest pieces of disinformation being propagated by the media. Let's say a discovery of 500 million barrels hits the headlines – goes round the world like wildfire. 'XYZ Company discovers 500 million barrel oilfield...' – and that's where it ends. The reporters, the news editors don't follow up on the discovery, because if they did, they would get a glimpse of the truer picture. The cost of producing oil from that well is basically the finding cost, plus the production cost, divided by the estimated numbers of barrels produced. Some wells run out at less than a dollar a barrel, others can cost many tens of dollars a barrel to produce. ANYTHING done to the well thereafter adds to the cost per barrel, and a lot has to be done to some reservoirs to keep them flowing, to extract the absolute maximum from them. Services and treatments can do wonders in improving production but at a price, and the oil company always has its eye on the bottom line, the total cost per barrel to produce the oil. When it become too unprofitable to produce, the well, the reservoir will be dumped like a hot brick.

But now, read this. It is from a 2004 report by Wood Mackenzie, a well respected oil industry analyst and consultancy company.

How much oil does the world have left?

PLENTY TO SHARE?: After the announcement by Anglo-Dutch oil giant Shell that its oil reserves are not as plentiful as they seemed, analysts reconsider their estimates:

How much oil do we have left? According to the United States Geological Survey's (USGS) latest report, published in 2000, the planet had 3 trillion barrels of oil and gas before we started using it up. It calculates we have used some 700 billion, leaving 2.3 trillion barrels underground. A simple calculation using data from the Center for Global Energy Studies (CGES) shows that with 28.8 billion barrels currently being used a year (79 million a day), there is some 80 years of supply left in the ground.

However, the 2.3 billion barrels left includes 1.4 billion, which, according to USGS analysis of global geology, exist but have yet to be discovered. That leaves roughly 890 billion barrels of oil and gas that have been discovered and are booked as proven reserves - roughly 31 years' supply.

This estimate is lower than the industry's. BP's oil statistics - the industry bible - indicate some 1,047 billion barrels of proved reserves - 36 years supply on current usage. But analysts don't worry about this too much. Derek Butter of consultancy Wood Mackenzie says: "It is not something we get hung up about. We think in the medium term there is enough oil in reserve."

The BP statistics show that over the past decade, reserves have been roughly constant - in 1992 there were 1,006 billion barrels - indicating that through the 1990s the industry has been effective at finding reserves to replace the resources we use. It has done well - in 1982, the figure was only 676 billion.

Manouchehr Takin of the CGES says: "Forty years ago, economists and geologists said there were 40 years of oil and gas left. Now they are saying the same." He adds: "The techniques for finding oil and extracting it are advancing rapidly. It shows that at any one time what is left, and the time it will take to use this up is only a snapshot dictated by the current state of knowledge. And, of course, there is the impact of economic development, and the efficiency with which we use energy."

Efficient energy use, he argues, has offset the growth of formerly non-industrialised countries, such as India and China, which have anyway not grown as rapidly as was expected in the Sixties and Seventies. Similar dynamics are likely to continue.

Nevertheless, demand for oil is likely to increase, as, over time, is the price. Former UK Environment Minister Michael Meacher pointed to a crunch by 2015, when he says demand is likely to total some 60 million barrels a day more than the current 79 million, while supplies will only be some 90 million.

CGES argues that Meacher has overstated the increase in demand. It argues that taking reserves into account, his estimates imply 4 percent growth a year, whereas, depending on price, CGES forecasts between 1 percent and 1.3 percent, while OPEC has 1.7 percent. At the top end of its range, CGES believes demand will be 87.7 million barrels a day, within its expected production level.

Nevertheless, as Wood Mackenzie points out, nothing can be taken for granted. In a note last week it said: "There is no escaping the fact that oil and gas are finite resources: the more that has been found, the less that remains to be found."

The crux is this: Will exploring and finding new reserves be worth it for companies? Theory suggests that as easy resources are depleted extraction becomes more difficult, but that reduction in supply increases price, underpinning the eventual return for companies. Wood Mackenzie is sceptical. It points out that between them major western countries need to find reserves equivalent to Angola's every 15 months just to stand still.

It shows that new areas such as Alaska face difficulties because of extreme environmental sensitivities and regulatory concerns. Other areas, such as Nigeria, and newer ones, such as Benin, are uncertain. Companies must take greater risks - such as exploration in ever deeper water, and expensive investment in extraction from shale beds. And it points out that reserve replacement has often been achieved by shuffling finds into the proved category rather than actually finding more oil (which underlines the importance of Shell's declassification of reserves).

Eventually, it will be down to consumers to decide when the price of oil and gas makes it no longer worthwhile investing in things that are powered by them. It is likely, concedes Takin, to be well before our notional 2.3 trillion barrels are used up ... whenever that may be.

In that respect, the world will never run out of oil. There will always be oil, in the reservoirs. The world will however run out of people willing to pay for that oil. A tank of gasoline may cost you US$ 60.00. That is still an acceptable price. But what if it cost US$ 600.00? Or more?

In a commercial sense, the price would have to be passed on down to the consumer – the people in the street. Your goods would become so expensive that they would stop buying - they would not be able to afford them! What we are doing right now is producing oil from existing reservoirs at a price competitive with other sources. Commercial forces are driving the cost of the oil and not true value. As said before, oil price is basically finding cost, plus recovery cost, plus processing cost. A barrel of oil from the huge Ghawar oilfield in Saudi Arabia, the single largest oilfield in the world, cost pennies to produce, but it sells at the OPEC-set price of between US$ 22.00 and 28.00*. On the other hand, the oil being produced from the Sakhalin Islands in Eastern Russia truly affords the stuff its name of 'Black Gold' - if not in value, at least in cost of production.

However, we are getting away from the point. Of course there is oil and gas, in abundant quantities, still available. The media lets everyone know about it. But we are using more than we can replace. Do the media tell everyone about that? Not likely. In the first place, gloom and doom doesn't sell newspapers or attract viewers to the TV channels. The majority of the population is happy to sit back and accept whatever its government tells it. Governments have to keep the country economy running, come hell or high water.

But we are living in a false economy. Everything is fine, except that we are bleeding to death. Your government, my government know about the supply situation but they are virtually powerless to do anything about it. Our world depends on energy, everything we do depends on energy. We are using more energy today than we ever did before in the history of Man. And our primary source is running out in a hurry.

It took many years to convince governments, industry that pumping CFCs (ChloroFluoroCarbons[†]) into the atmosphere was destroying the ozone[‡] layer, a layer of gas at the top of the atmosphere (about 30 - 40 km straight up) that is essential to life on Earth. Ozone, a molecule of three oxygen atoms is poisonous at ground level, but a life saver at 40 kilometres, as

* …today's price, heading for $150 a barrel, these figures are rather a farce, but they were set by OPEC in an attempt to reassure the world that they could control production and hence price. If oil went over &28.00 a barrel, OPEC would increase production to stabilise price and bring it back into the $22-$28 range. It has never worked. Nor, for that matter, has OPEC.

†Organic compounds that contain carbon, chlorine, and fluorine atoms. CFCs are highly effective refrigerants that were developed in response to the pressing need to eliminate toxic and flammable substances, such as sulphur dioxide and ammonia, in refrigeration units and air conditioners. They were also widely used as aerosol propellants, cleansing agents for electrical and electronic components, and foaming agents in shipping-plastics manufacturing.

‡Ozone is vital to human and animal survival because it is responsible for the absorption of the sun's ultraviolet light. Without this protection, blindness and skin cancers could result from penetrating ultraviolet light. In 1987 an international treaty, the Montreal Protocol, called for reducing CFC use by 50% by 2000. A 1992 amendment to the treaty called for the end of CFC production in industrial countries by 1996, and by 1993 CFC emissions had dropped dramatically.

it reacts with and absorbs the energy from the incoming ultraviolet light from the sun. Eventually, as the general population began to realise the direct threat to their lives, governments were pressured into forcing not only the reduction in CFC usage, but the complete abolishment of CFCs for general use.

Although other molecules, other compounds that we pump into the atmosphere also have a deleterious effect on the ozone layer, with the banishing of CFCs, the ozone layer is slowly healing itself, and should eventually revert to its original condition, as a shield against ultraviolet light.

That's the funny thing about the Earth. Most of what we do to it is on the surface. We can scar it, change its chemical complement, tear down and burn its vegetation etc., the Earth will survive. Changed perhaps, but it will survive. This is one of the aspects of living on Earth that our ancestors took very much for granted. They saw how the Earth could absorb such effluent and still come up shining the next morning. Think of the industrial revolution, huge smokestacks in Europe belching black, sooty smoke into the atmosphere. As cities were built, waste was simply pumped into the sea, the great oceans could absorb it all. The population needed energy and it came from coal (mostly) and timber. Both irreplaceable (at least in the short term) sources of energy.

And that was on a population base of 500 or 600 million people! Today, we have a population 10 times that, which suggests ten times as much effluent, ten times as much industrial waste, ten times as much energy usage.

But it is much, much more than that. We have become a throw-away society. Look around you at the landfills, the need for space to create landfills. You can build an incinerator to get rid of it, but not in my backyard. There is so much waste to get ride of that it is a real and permanent problem to our society. We ship unwanted waste to other countries – let their people worry about it. This is not a solution. Nuclear waste especially, is a major problem, but we'll come back to that in a later chapter.

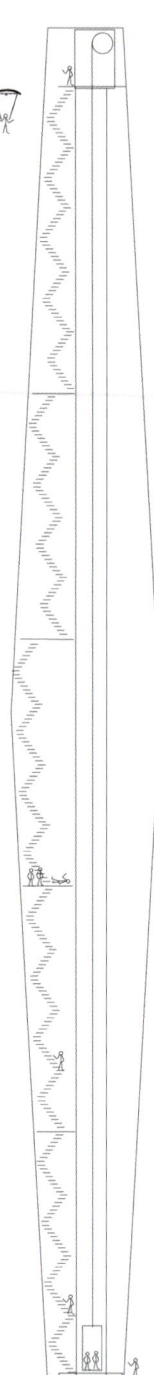

The largest jump has been in energy usage. Our forebears built enormous cities, beautiful cities of elegant buildings with wide, spiralling staircases – that's right staircases. No building was so high that you couldn't reach the top except by staircases.

Lifts, or elevators did not come into use until electricity ceased being a novelty and became a supplied service, although steam powered 'lifts' were in use in British mills by the turn of the nineteenth century. The first real elevator, the invention that would change our city skylines forever, was installed by Elisha Graves Otis in New York in 1857, although it was not until 1889 that his sons, the Otis Brothers, revealed the first successful direct-connected geared electric elevator machines. Merging with other elevator manufacturers to form the Otis Elevator Company, in 1903 they introduced the design that would become the backbone of the elevator industry, the gearless traction electric elevator, '…engineered and proven to outlast the building itself…'.

This opened up the way for city architects to build upwards - already land was becoming a problem - instead of outwards. Look around you today, look at the condominiums, the office blocks, the hotels, the corporate headquarters – all cleaving the sky, not one of them being able to exist without the elevator. But the cost in energy consumption is enormous. The elevators, the lighting, the air-conditioning – all consuming huge amounts of energy.

Cities are enormous consumers of energy. Cities attract people looking for work. They need to, to survive and grow. In the USA less than two percent of the population is employed directly in agriculture – where once upon a time Man's was a 100 percent agrarian society, now that society is reduced to two percent of the whole population. The USA has one of the most efficient agricultural systems in the world, where so few people can provide for a nation of more than 290 million people. It is a tribute to the mastery of agriculture by Man, that America can feed not only its own population but provide sufficient surplus to feed half the world!

But that very efficiency also creates employment problems for the rest of the population. When you are fed, housed, clothed – what else do you need? Our agrarian society has changed. It has become a much more complex society. People find employment in providing the food to feed such a huge population. There are many more people involved in food preparation, than there are in primary production. People find employment in building mass housing projects, whether a whole new town, or high rise condominiums. And people find employment in making the clothes that the rest of us put on our back. A society not much different to the one that evolved when people first began to congregate in villages all those centuries ago.

Even considering only those three basic requirements of survival, the demand for energy soon becomes enormous as people move into city dwelling. Energy is needed for transport: the trains, trams, buses, private and goods vehicles needed to get from place of abode to place of work, and back again as well as to distribute the manufactured goods. Energy is needed for lighting, for escalators and elevators to get around these enormous buildings. Energy is needed to run the factories, the offices the workplaces that employ these people. And people themselves use a tremendous amount of energy – to run their TVs, refrigerators, air-conditioning, cookers and ovens, lighting, computers, bed blankets and electric toothbrushes.

And to produce all of these, the TVs, refrigerators etc., we need more people. More work, more people, more factories, more energy needed. It's a continuously growing pattern. People are beginning to realise the stresses and strains of city life and are moving out, to a private hilltop, where they can afford it, although the majority of them are stuck where they are, to face the problems of city life. Such as the massive blackouts that hit the eastern seaboard of the USA. Energy demand was simply too great for the system to handle and everything came to a standstill.

This is only an example, but it is an example that will be recurring more and more as time goes by. Our ever-growing population is putting a tremendous strain on our energy supply - not only on the energy supply, but on the energy distribution channels. We can improve and renew the distribution channels, but much of the energy supply is not renewable or replaceable.

And what worries many people nowadays is the Catch 22[12]* situation into which we have gotten ourselves. A perfect example is the city of Seattle in North America. A sleepy little township turned into a bustling metropolis by one product – the aeroplane. More specifically, Boeing aeroplanes. Boeing set up its manufacturing facility in the town of Seattle and today, along with Microsoft, employs the majority of the 600,000 citizens of that township. Boeing has had a prolific, if chequered history in the aviation industry but the name of Seattle is today worldwide synonymous with Boeing. So when Airbus Industries of Europe overtook Boeing in the high flier stakes, Boeing was, to say the least, a little put out. After being industry leader almost since the inception of flight, Boeing was relegated to second place. To regain its lost crown, Boeing has pushed forward the design and delivery of the latest and greatest model to come off the design boards at its Seattle plant, the Boeing 7E7, an aircraft that is every bit a 21st century machine. Designed to come into service by 2008, the plane uses up to 20 percent less fuel than other jets of equivalent size and range.

Let's play that back... the plane uses up to 20 percent less fuel than other jets...

Isn't that a saving? Isn't that good news? Where's the Catch 22? Well, the plane will use, unless a revolutionary new drive system is invented within the next six months or so, basically the same turbofan engines

---

*A situation or predicament characterized by absurdity or senselessness... A contradictory or self-defeating course of action.

as any other plane, which burn hydrocarbons for power. A Boeing 747, with four Rolls Royce, Pratt & Whitney or General Electric turbofan engines will consume somewhere around 56,000 gallons of kerosene in a long-haul flight. So even if the 7E7 (now renamed the 787) uses 20% less, it will still need 45,000 gallons of fuel. Boeing expects to build between 2000 and 3000 of them over the next twenty years. Let's say they build 2000. That's 90 million gallons of fuel needed each time they all take off for a long-haul flight.

Planes like these, even if superbly efficient compared to planes of yesteryear, consume enormous quantities of hydrocarbons to carry people around the world - people going on holiday, going to study, going on business...

To all intents and purposes a gallon of kerosene equals a gallon of crude, so to fly these planes we will need about (42 gallons to a barrel) 2 million barrels of crude oil each time they all take off. These planes are designed for a minimum 20 year lifespan (there are 747s from the first batch in the early 1970s still flying, over 30 years now) and will need the same kind of fuel over the 20 to 30 year lifespan. Where are they going to get it? We're not talking about just Boeing 7E7s either. We're talking about Airbus A380s, the huge 500 seat planes that finally defeated Boeing. We're talking about the planes already in services. Of the 86-odd million barrels a day of crude that we are consuming a large proportion of it goes to aviation fuel. Even if an engineer somewhere in the world ran out of his laboratory tomorrow yelling "Eureka!! I have solved the problem…", waving plans for a revolutionary new motive force that would provide the power to get aeroplanes off the ground and into the air, it would take at least another ten years to refine, design, build, test and retest until it passed every certification requirement of the aviation industry. So the companies and the governments responsible for these planes being built in Seattle and elsewhere are banking on sufficient fuel being available for the next ten, twenty or thirty years!

What alternative do they have? "Sorry folks. We know that in ten years from now, there will be insufficient fuel to keep fleets of aircraft aloft as we do now, so we might as well save our money, stop wasting our precious resources, shut up shop and all go back to farming."? A city like Seattle would self destruct. Of course, they could phase it out slowly, let people retire naturally, just don't hire on any new staff. It won't make any difference. The end result will be the same. Just as the supply curve for the energy source – hydrocarbons – will be tailored to our behaviour. Keep on using energy at the present rate and the beginning of the end will be hastened. Or we can reduce our consumption to make it last as long as possible, but in the final result, the end will be the same. We are all caught in our own Catch 22 situation. Or to put it another way, we are damned if we do and we're damned if we don't. (See Appendix VI)

"Water, water, every where, And all the boards did shrink; Water, water, every where, Nor any drop to drink. The very deep did rot : O Christ ! That ever this should be ! Yea, slimy things did crawl with legs Upon the slimy sea. About, about, in reel and rout The death-fires danced at night; The water, like a witch's oils, Burnt green, and blue and white."

- Samuel Taylor Coleridge, the Rime of the Ancient Mariner.

05

WE CAN'T HAVE
IT ALL

Having been shot by the Ancient Mariner, the Albatross wreaks its revenge. We may have already shot our own Albatross.

A very famous song by American singer Lynn Anderson starts off with 'I never promised you a rose garden…'. Sounds a little prophetic, but every bit as meaningful in whatever context you wish to consider it – whether it is love between a man and a woman; an employer's relationship to his employees; or a government and its people – you can't have it all. Nobody, none of us, was born into a society that would give us everything, whatever we wanted, take care of our every need until the day we die. We cannot expect to live our lives out in a rose garden.

And yet, a great majority of us think that way, 'Why can't the government do this?', 'What is the government doing about that?'. We expect to be cared for, catered to by our governments, our governments should handle all our problems, our governments should ensure we get whatever it is we need to live – food, water – energy. Some well meaning, responsible governments do just that – bend over backwards to try and provide the best for their citizens. But it all costs money, and most people don't want to end up footing the bill. One day, there will be no option and the problems, and costs, being faced right now will seem as nothing.

As has been stated before, the greatest proportion of crude goes to providing the fuel for transport. Since there are few alternatives to gasoline, kerosene or diesel powered engines, this will be the major problem that will have to be faced in the future. Coal, and some oils, go to driving power generation equipment, largely huge, smoke-belching monstrosities that sit on the edge of your cities or, in some cases, smack in the middle of your cities. These plants utilise coal-fired, or oil-fired furnaces to heat water to create steam to drive turbines, to drive huge power generators.

You can almost see the inefficiency of it. Coal to heat, very inefficient. Heat to steam? Not too efficient either but more efficient than coal to heat. So that is one obvious area to be looked at more carefully. Steam to rotation, through the turbines, relatively efficient compared to the previous steps. Rotation to electricity? Quite a lot of losses here as well, so all in all, converting coal to electricity is not very efficient at all. As has been said before, we are a wasteful bunch. Using oil-fired burners does not improve our lot very much, we still lose a lot of the original energy in place to heat loss to the atmosphere, conduction through the many pipes and pieces of equipment that need to be kept hot, friction in the turbines and generators, etc., etc.

A lot has been done to improve such plants, and the efficiency has risen considerably. Newer plants now use natural gas and are highly efficient, with cogeneration ability, using the heat generated by the high pressure gas turbines to make steam to drive low pressure steam turbines. Such co-generation plants can add 20 percent to the overall efficiency of the energy conversion.

There are of course many other means of creating electricity, with governments investing varying sums of money in one or more means, depending on immediate need and foresight*. Many, still in the experimental stage, may never get off the ground, while others will

---

*In an excerpt from a report in 2004 by Cambridge Energy Research Associates (CERA) the authors suggest that blistering economic growth in China, and its surge effect on demand for energy (along with most other commodities), have pulled international energy markets onto the tiger's back, the tiger in this case being China. An old Chinese proverb says 'When riding a tiger, it can be difficult to get off'. This is particularly appropriate when analysing the economic boom-driven energy supply quandaries facing China today.

CERA Chairman Daniel Yergin (of 'The Prize' fame) and Scott Roberts, CERA's Director of China Energy, in Riding the Tiger: The Global Impact of China's Energy Quandary said that between 2000 and 2004, China accounted for an astounding 40 percent of the total growth in world oil demand, and its rapid economic growth has turned it into the world's second largest oil market. In a decade China has gone from self-sufficiency to being the most dynamic factor in the world oil market and one of the main elements in today's oil price. This is in an environment in which 24 of 31 provinces report electric power shortages, and blackouts are common in industrial centres, where average oil refinery utilisation is 92 percent and Yangtze River oil depots and bargeways suffer traffic bottlenecks.

provide at least some of the energy needed to continue. Let us now take a look at some of these alternative energy sources.

One of the oldest and most ubiquitous is water power. Hydroelectric power is an indispensable source in countries with the requisite conditions: high rainfall in mountainous areas; powerful, high-volume flowing rivers; suitable terrain to build dams, such as the Three Gorges Dam at Sandouping, Yichang, Hubei province in China. The Three Gorges Dam is approximately 180 metres high and is expected to be in full production by 2009 when all 26 water turbines will be operational, producing some 18.1 million kilowatts of electricity. Built, at a cost of around US$ 25 billion (some say as much as three times that figure), on the Yangtze River, the project displaced more than a million people as the floodwaters rose. When full, the dam will hold some 1,377,285,000,000 cubic feet (39 billion cubic metres) of water, be over 520 feet deep and almost 40 miles long. (In contrast, an Olympic swimming pool holds about 137,700 cubic feet of water).

China is estimated to need nearly $1.4 trillion in energy investment during the first 20 years of this century, an amount equal to the regional totals of Latin America and Africa combined. However, capital investment is hampered by structural problems in the economy. "The pace and direction of China's short- to medium-term development is less certain – and far more volatile – than most observers now anticipate. Indeed, there is the possibility within the next several years of a more significant slowing of growth, or worse: a so-called hard landing," Yergin and Roberts observed.

"Dismounting from a tiger is never easy, and there remains the possibility of a hard landing. This result would have a sharply corrective effect on industry and would directly affect energy demand – including oil imports. A hard landing scenario, therefore, could be a surprising event for international oil markets that are now betting on straight-line growth. This turn in trend, taking place amid strong gains in non-OPEC oil production and expansion of oil production capacity in some OPEC members, could have a chilling effect on global oil prices. Although a price collapse is far from certain if China hits a rough patch, OPEC cohesion would be put to the test."

China will overtake the US before 2010, according to the World Energy Outlook, as the largest energy consumer and, while China's energy resources, especially coal, are extensive, they will be insufficient to meet the growth in energy demands. China needs an additional 1300 GW of electricity generating capacity – more than the total current installed generating capacity of the United States!

Not only that, India, with its unprecedented growth plans for 96% of the population having access to electricity by 2030, hopes to meet these demands through imports! By 2025 or before, India will overtake Japan, to become the third largest importer of oil, after China and the US. By 2030, it is projected that India's additional generating capacity will rise to 400 GW, more than the combined capacity of Japan, Korea and Australia.

It is a marvellous human achievement and will alleviate China's pressing energy problems but, building such dams comes at a price. Sometimes a terrible price. Some years ago, earthquakes in a relatively stable part of Eastern Siberia puzzled Russian scientists and geologists, until it dawned on them that the weight of water accumulating in a newly dammed lake thousands of miles to the west, was tilting the crustal plate that Siberia sat on sufficiently to induce earthquakes.

Just to illustrate the potential for disaster that dams such as the Three Gorges Dam pose, this article* highlights the kind of situation we could end up facing. Even though this dam was created naturally, the forces and the results would be very much the same.

On an icy February night in 1911, an earthquake rocked Tajikistan's remote Pamir Mountains. Landslides of mud and rock tumbled down the foothills, blocking the flow of the River Murgab and forming a natural dam and lake within minutes of the powerful tremor. Now, almost a century later, geologists warn that the slightest shudder in the earthquake-prone mountains could cause Sarez Lake to rupture the natural dam that contains it, unleashing a wall of water that would flood hundreds of villages in Tajikistan, Uzbekistan, Turkmenistan and Afghanistan.

*1999, Cynthia Long, Staff Writer at DisasterRelief.org

"Millions of tons of water in Sarez Lake hang over the surrounding countryside at an altitude of over 3,000 meters (nearly 10,000 feet)," said Tajik scientist Oleg Barotov. "It is truly hanging because the Muzkol and Northern Alichursky mountain ranges in which the lake is situated are subject to considerable seismic activity and frequent earthquakes."

High, frigid and remote, the Pamir Mountains contain several ranges which are primarily located in Tajikistan but also stretch up to Kyrgyzstan, down into Afghanistan and east to China. In fact, the name Pamir itself refers to the great valleys that divide the various mountain chains that criss-cross this rugged swath of the former Soviet Union.

Often described as the "Roof of the World," the Pamirs are the highest upland in the former Soviet republic and their tallest points -- with names like Peak Communism and Peak Lenin -- are reminiscent of the communist regime. Though the mountains have attracted climbers and hunters for years, the region has always been seismically volatile -- at least three dozen earthquakes have rattled the mountains in the past decade alone.

The area's seismic activity alarmed Russia's Tsarist officials who hastily set up a lake observation station in 1913, two years after the dam was formed. But it was not until 1934 that researchers began to recognize the danger of erosion and the possible collapse of the Usoi Dam. In the 1950s scientists and engineers started monitoring the stability of the natural dam and calculating theoretical measurements of the gigantic flood wave that would charge down the valley if the dam were to burst. Finally, Russian officials proposed reducing the lake waters by using it for hydroelectric power production, but the proposal collapsed along with the Soviet Union and the current independent government of Tajikistan.

Concerned about an impending flood, an international team of geologists and aid workers visited the region early this June and predicted that even the slightest "seismic shiver" could breach the mountaintop dam. Not only is it already leaking, an unstable cliff with deep cracks is hanging over the north bank of the lake.

"In the event of an earthquake, it is likely that the cliff would fall into the lake," said a recent report commissioned by the UN's International Decade for Disaster Reduction. "A rush of water would strike the dam and cascade over the top." Such a powerful surge could severely damage or destroy the weakening dam.

If the dam ruptured, the team predicts that a wall of water more than 300 feet high would rush down the narrow Murgab River gorge into the Bartang valley, destroying hundreds of mountain villages and flooding a 20,000 square-mile area inhabited by 5 million people. According to Scott Weber, the UN Office for the Coordination of Humanitarian Affairs delegate who coordinated the expedition to the lake, the wall of water would still be as high as a two-story house more than 600 miles downstream.

Lake Sarez, deep in the Pamir mountains of Tajikistan, was created when a strong earthquake triggered a massive landslide that, in turn, became a huge dam (west end of lake in main image and centre right in insert) along the Murghob River, now called the Usoi Dam. The resulting lake is perched above surrounding drainages at an elevation greater than 3000m, and is part of the watershed that drains the towering Akademi Nauk Range. The lake is 61 km long and as deep as 500 m, and holds an estimated 17 cubic km of water. - http://earthobservatory.nasa.gov

Weber told New Scientist magazine that strengthening the dam is impractical -- it is over 1,500 feet high, nearly two and a half miles long and is located in one of the most remote and rugged areas of the world. And tinkering with the colossal structure could upset its already fragile stability. The team is in favor of a plan put forward by the Tajikistan government to siphon water from the Sarez and flow it down the valley through natural and manmade channels to the drought-stricken Aral Sea region.

But for the villagers living downstream, the immediate need is for safe havens and an early-warning system. Residents had previously relied on an automatic radio notification system which signalled by satellite any movement in the dam, but funding for the system fell with the Soviet Union.

Though necessary to avert a major disaster, an effective warning system and the proposed pipelines to siphon lake water will require huge investments, not only from Central Asian states but also from the international community…

As can be seen from the above, the prospects for a human disaster virtually unheard of before are very real with the Three Gorges Dam. The Yangtze River is renowned for its flood plain, which in actual fact was one of the reasons given for building the dam – the ability to control the flooding of the river and the subsequent drowning of so many people over the years (over one million in the past 100 years). But the dam was built with less than total agreement amongst the people of China, it was more political will that built the dam than engineering savvy and expertise. Many people, even Chinese engineers, have spoken out about the quality of construction. Typing 'Three Gorges Dam' into an Internet search engine will provide the reader with an almost unending list of material, for and against the building of the dam. Unfortunately, mostly against. It isn't just the quality of the construction that is in question either, it is the potential for a disaster beyond our imagination (some 400 million people live below the dam) that bothers engineers, scientists and others throughout the world. The Yangtze is infamous for its heavy cargo of silt that it brings to the eastern regions. With the dam slowing the flow, much of this silt will be dropped out above the dam. It would not be the first time that a dam has been breached as the waters overtook the dump sluices, because of silt build-up at the base of the dam.

So, building a large hydro-electric scheme to provide power is not a simple process. Having gone ahead with the Three Gorges Dam, China faces at least one other problem it didn't need to contend with, the possibility of inducing earthquakes as the weight of water causes the earth to shift its balance. And should one of these earthquakes damage the dam, we can only imagine what the end results might be.

There are so many water-based power generation possibilities today, some of them reaching back into antiquity and the days of the Greeks, who were, after all, master craftsmen who discovered and invented a wide range of things. Although details of its use can be traced back to 4000 BC, in all probability one of their inventions was the waterwheel, a very simple device which remained unchanged over centuries and which still finds use today in some way or another. Consisting of a wheel with buckets or paddles, it was placed such that a stream of water flowed under it, or via an aqueduct, over it, thereby turning the wheel. Depending on the size of the wheel, the volume of water flowing through it and the constancy of the water flow, the rotation of the wheel was put to a variety of uses – grinding corn for instance.

The waterwheel frequently became the centre of a town, as it was the only source of power until steam engines were invented. They could be used to drive sawmills, to saw up timber to build houses; they were used to operate pumps, to lift water for field irrigation, or drainage and such like. They also played a big part in providing energy to textile mills at the beginning of the Industrial Revolution, but their biggest debt is owed by the metalworking industry, where the power from that one rotating shaft drove things like trip hammers, bellows (for the forge), shears and all manner of machines. I still remember the fascination, as a child, of being in one of these places, (although by then driven by a steam engine nothing else had changed), with criss-crossed belts flap, flap, flapping, spindles turning spindles, machinery thumping, crunching, grinding or simply cycling quietly – I had no time to be scared, it was so fascinating to see.

Sad really to think of the demise of such marvellous, ingenious inventions* that had to give way to efficiency and cold, hard, machined steel. Waterwheels were too slow to ever be used except in the most rudimentary form to generate electricity, so their usefulness was limited to the immediate vicinity of the mill itself, heralding their own end as more and more energy was demanded.

We will come back later to look at the ways in which water can be utilised to create energy.

---

*Anyone visiting the gold town of Sovereign Hill at Ballarat, Victoria, in Australia can still see such a system in operation. It is maintained as part of a working museum piece, an 1850s goldfields township where real gold flows in the creek! The workshop produces a variety of useful household metal objects.

"It is given to but few men to achieve immortality, still less to achieve Olympian rank, during their own lifetime. Lord Rutherford achieved both. In a generation that witnessed one of the greatest revolutions in the entire history of science he was universally acknowledged as the leading explorer of the vast infinitely complex universe within the atom, a universe that he was first to penetrate."

The eulogy in the New York Times, 1937, to Ernest Rutherford, who discovered the structure of the atom.[12]

06

THE ALTERNATIVES AVAILABLE

In an earlier chapter we touched briefly on the subject of atomic, or nuclear power. If you remember, it was stated that in the conversion of chemical energy to some other, more useful form of energy, only the electrons in the outer shell of the atoms participated in the reaction. In reality, this should be called atomic energy, as it is the energy of the atoms and molecules themselves that is being exploited. In the case of nuclear energy, it is the nucleus, or core of the atom that is important in energy conversion.

Let us imagine a bowl full of identical red and yellow balls, except that the red balls have a positive electric charge. Positive repels positive, so to keep the red balls in the bowl from flying apart, we put in the yellow balls, which will mix with the red balls and, in very close contact, will 'stick' to them. The more red balls we put in the bowl, the more likely they are to come close to another red ball, so we have to add more and more yellow balls to keep them apart.

The red balls are the protons and the yellow balls are the neutrons that go to make up the nucleus of the atom. If you can think back to basic chemistry, you will remember that it is the protons that decide what an atom is. Neutrons, although they play an essential role in keeping an atom stable, are really only passengers in the nucleus. The 'glue' that holds them all together is the all important substance in nuclear energy.

Hydrogen is the simplest atom, having only a single proton as its nucleus. Helium, next in the Periodic Table, has two protons and two neutrons. And so it goes, the addition of each proton, in most cases, requiring the addition of one or more neutrons. If you look at a copy of The Periodic Table (Appendix IV) you will see several numbers on the individual blocks. 'Z' is the Atomic Number, i.e., the number of protons in an atom. 'A' is the Atomic Weight, the combined mass of the protons and neutrons, which may not necessarily be a whole number - any guesses why it may not be a whole number? The answer lies in the fact that it is an average weight, and many atoms have isotopes, or nuclei with differing numbers of neutrons.

06

This difference in atomic weight was one of the discoveries that led to the making of the atomic bomb. Elements in the lower reaches of the Periodic Table are considered to be stable, i.e., if certain atoms were carbon 500 million years ago, chances are they are still carbon atoms today. However, as the element gets heavier, it becomes more and more difficult for the neutrons to hold the nucleus together, until a point is reached where the nucleus will disintegrate. That point is Uranium. Elements after uranium are unstable and will eventually 'decay' or disintegrate into other elements. All elements after uranium have never been found in nature because they had all decayed into base products long before Man took his first steps on Earth. They are 'manufactured' in laboratories.

Uranium was first discovered in 1789 by a M. H. Klaproth, in Berlin, but it was only in 1842, in Paris, that E. M. Peligot finally isolated the element. Named after the planet Uranus, it is a radioactive, silvery metal consisting of more than 99% uranium-238, traces of uranium-234 and less than 1% uranium-235. This rare form, U235 or ^{235}U as it is more correctly written, is the one needed for nuclear energy. To separate the two forms, uranium must first be turned into uranium hexafluoride, a volatile liquid. Then, by using the different weights of the atoms, the ^{235}U is slowly 'enriched' by increasing the concentration of these atoms. It is a slow and expensive process. World production of uranium is 35,000 tonnes, with reserves amounting to over 200 years supply.

A uranium atom will spontaneously disintegrate into lighter weight elements, such as lead. But there is no specific 'time' at which it will do this, so an 'average' is measured, at which time in any given volume of uranium half of the atoms will have disintegrated. This is called the 'half-life' of the element. Uranium-234 has a half life of 244 thousand years, but only constitutes about 55 atoms out of every million atoms of uranium. Uranium-235 has a half life of 704 million years, and makes up roughly 7,200 of the million atoms. Uranium-238, which makes up the remaining 992,800 atoms in our million, on the other hand has a half life of 4.5 billion years. Since that is the same age as the Earth, we can safely say that ^{238}U is not radioactive.

Or can we? Remember, we are talking half life. Not age. So in the history of this planet, ^{238}U has lost roughly half of the original mass of uranium that was part of the agglomeration of matter that became the Earth.

Another element, this one man-made, is called Lawrencium and has a half-life of 180 seconds. That is, every 180 seconds half of the existing atoms decay into something else. Are we likely to find any around today from the original mass of the Earth? I very much doubt it!

But what has this atomic weight and half-lives got to do with energy? In a word, lots! As already mentioned, some elements can be found as isotopes, uranium being one of them. Uranium has three isotopes, ^{234}U, ^{235}U and ^{238}U, in other words, some atoms have three or four more neutrons (or three or four less neutrons) than others. Why does it have to be neutrons? If it were three or four more protons or three or four less protons, it wouldn't be uranium, right?

By measuring the number of each type of atom in a known sample, scientists were able to figure out that ^{234}U and ^{235}U were the more unstable atoms. Although ^{234}U is even more radioactively unstable than ^{235}U, since it appears in such limited quantity, it plays small part in energy equations.

Scientists had already discovered that during its disintegration, a uranium atom gave off energy. What it was, they were not sure, but at the time they were working with high energy X-rays, which exposed photosensitive film, and because they knew that X-rays were a form of radiation, they called this new energy 'atomic radiation'. Any element that exhibited such effect was called radio-active, a name that stuck. Through the course of their experiments, scientists soon found that there were three distinctly different types of 'radiation' emanating from elements that exhibited this 'radioactivity'. These they named Alpha (given the Greek symbol α), Beta (Greek symbol β) and Gamma, (symbol γ). Alpha was a fairly weak 'ray' being stopped by something as thin as a sheet of paper or a few inches of air. Beta 'rays' on the other hand were much more energetic while Gamma 'rays' were the most energetic, requiring several inches of lead to stop them! They were even more energetic than the X-rays being used in the original experiments!

A little bit of simple mathematics however, showed that if you added up the atomic weights of all the daughter elements of ^{235}U, the decay elements it broke down into, it didn't tally. Something was amiss. There was obviously more to radioactive decay than the release of energy. It transpired that the decay from uranium to lead (see chart) was a complex path, in which all three types of radiation – alpha α, beta β and gamma γ – were involved. Alpha radiation is a particle, basically a helium nucleus, while beta radiation is a high energy electron. Gamma rays are the radiation that first attracted Becquerel* by leaving an imprint of the piece of pitchblende on his photographic plate. Gamma radiation is more powerful even than X-rays and the one that does the damage in radiation exposure.

However, the beta radiation was the most interesting, as far as our search for energy is concerned. As stated, beta radiation is high energy electrons. Where did they come from, if we remember that electrons form clouds around atoms? And what could eject them with such force? Free electrons will wander from cloud to cloud, but these electrons were exiting the atom in a real hurry. The answer lay in the expulsion of a positive charge, a positron[†], from a proton as the nucleus tried to achieve stability. But what was left behind? A neutron of course, with just about as much energy as the departing positron, only, being neutral, scientists had no way of detecting it at that time.

---

*Henri Becquerel (1852-1908), a French scientist, was studying the ability of uranium compounds to give off visible light after being exposed to sunlight (fluorescence). In 1896, he discovered by accident that uranium ore gave off certain invisible rays without exposure to the sun. These rays could penetrate the light proof covering of a photographic plate and affect the film just as if it were exposed directly to light rays. Becquerel referred to these invisible rays as radioactivity. In 1903, Becquerel shared the Nobel Prize in physics with Marie and Pierre Curie for their work on radioactivity.

[†]A positron is a positively charged electron, as opposed to a normal electron, which is negative. Positrons spontaneously annihilate as they collide with electrons, giving off, guess what? Energy.

To cut a long story short, these ejecta, the positron and neutron, have sufficient energy to impinge upon - gate crash more like it - the tranquillity of another atom's nucleus, let's say a ^{238}U atom. Should a neutron get through to the nucleus, because it is chargeless, it gets absorbed into the nucleus, becoming part of it. But ^{239}U doesn't exist. It is unstable. What happens is that the nucleus forces one of its neutrons to dump an electron, or Beta particle, raising the Atomic Number by one, to Plutonium 239. A quick mathematical sum will show you that if ^{239}Pu (94 protons) were to eject one Alpha particle, we end up with ^{235}U (92 protons).

This chart is extracted from the Nuffield Foundation's Book of Data (1972), editor R.D. Harrison – a 'must have' book for any seriously minded science student.

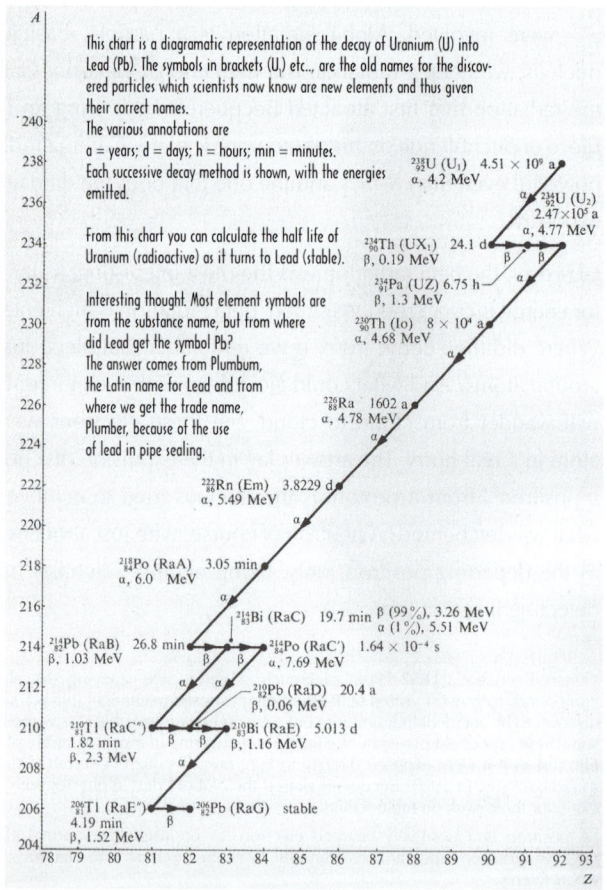

This chart is a diagramatic representation of the decay of Uranium (U) into Lead (Pb). The symbols in brackets (U₁) etc.., are the old names for the discovered particles which scientists now know are new elements and thus given their correct names.

The various annotations are

a = years; d = days; h = hours; min = minutes.
Each successive decay method is shown, with the energies emitted.

From this chart you can calculate the half life of Uranium (radioactive) as it turns to Lead (stable).

Interesting thought. Most element symbols are from the substance name, but from where did Lead get the symbol Pb?
The answer comes from Plumbum, the Latin name for Lead and from where we get the trade name, Plumber, because of the use of lead in pipe sealing.

$^{238}_{92}U\ (U_1)$ 4.51×10^9 a
a, 4.2 MeV

$^{234}_{92}U\ (U_2)$
2.47×10^5 a
a, 4.77 MeV

$^{234}_{90}Th\ (UX_1)$ 24.1 d
β, 0.19 MeV

$^{234}_{91}Pa\ (UZ)$ 6.75 h
β, 1.3 MeV

$^{230}_{90}Th\ (Io)$ 8×10^4 a
a, 4.68 MeV

$^{226}_{88}Ra$ 1602 a
a, 4.78 MeV

$^{222}_{86}Rn\ (Em)$ 3.8229 d
a, 5.49 MeV

$^{218}_{84}Po\ (RaA)$ 3.05 min
a, 6.0 MeV

$^{214}_{83}Bi\ (RaC)$ 19.7 min   β (99%), 3.26 MeV
a (1%), 5.51 MeV

$^{214}_{82}Pb\ (RaB)$ 26.8 min
β, 1.03 MeV

$^{214}_{84}Po\ (RaC')$ 1.64×10^{-4} s
a, 7.69 MeV

$^{210}_{82}Pb\ (RaD)$ 20.4 a
β, 0.06 MeV

$^{210}_{81}Tl\ (RaC'')$
1.82 min
β, 2.3 MeV

$^{210}_{83}Bi\ (RaE)$ 5.013 d
β, 1.16 MeV

$^{206}_{81}Tl\ (RaE'')$
4.19 min
β, 1.52 MeV

$^{206}_{82}Pb\ (RaG)$ stable

A

78 79 80 81 82 83 84 85 86 87 88 89 90 91 92 93
Z

This process of change and decay releases infinitesimal amounts of energy, as positrons are released, neutrons captured, protons break apart, which on its own is worthless. It is the ability to sustain activity, to have enough neutrons careening around to initiate a self sustaining reaction that becomes hugely useful in nuclear energy. An uncontrolled reaction would eventuate in an explosion, as witnessed so often by now in the bombs that devastated Hiroshima and Nagasaki in Japan, whereas a controlled reaction can provide heat energy to generate electricity, or direct electrical energy to power small scale requirements such as on board a spacecraft. Can you see why nuclear energy could not be used to provide large scale direct electricity generation? Like fire before it, nuclear energy is a hard-working slave but a terrible master.

One of the drawbacks of nuclear energy is the environmental (and psychological) impact on the population. Especially when it comes to Plutonium and its association with 'dirty' bombs, country-wide pollution and its ease of making. Not long after the end of the Second World War, a reactor was built at Dounreay, in the far north of Scotland, which was a fast breeder reactor, and which was used as an experimental station for most of its life. Even today, controversy still rages over the station, long after it has been shut down, after many years of accident-free, safe operation.

Everyone agrees that nuclear energy as a day to day power source is an almost inevitability, but the general opinion is, "Not in my backyard". There are many power plants already in operation, many more will need to be built. A recent report (UK Government White Paper & others on nuclear energy in Europe) gives some idea of how dependent we will become on nuclear energy. While Romania depends on a mere 10% of nuclear power, the Ukraine relies on nuclear energy for more than 50% of its supply. France depends on nuclear power for almost 80% of its electricity!

There are currently about 440 reactors in use around the world according to 'Forbes Asia' with another 30 under construction, 86 planned and

223 proposed. Uranium, like oil is a non-renewable commoditiy and subject to the wild swings in cost that demand and supply bring. All the same, global consumption is running at around 80,000 tonnes a year with supply averaging 42,000 tonnes. The balance is being made up from weapons-grade uranium which will disappear by 2013. Major new mining discoveries will have to be made if nuclear fuel is going to be a viable energy source beyond the next 30 years.

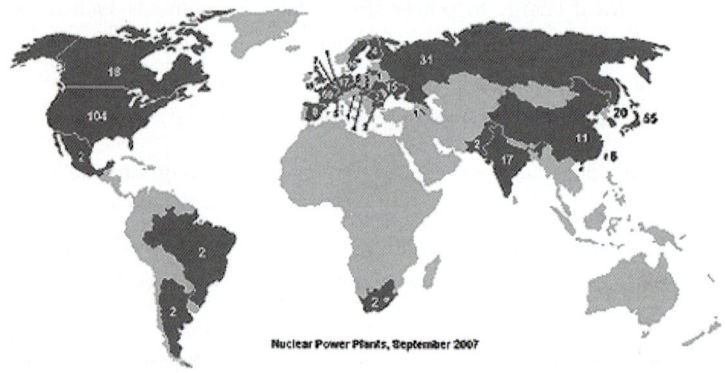

Nuclear Power Plants, September 2007

Where they are (IAEA-2007)

"Native Americans have called the area that was to become **Yellowstone National Park** 'home' for over 10,000 years."

History of Yellowstone National Park. Almost definitely amongst the first users of geothermal energy.

07

USING WHAT WE HAVE

A feature of this Earth that we do not make full use of is its geothermal potential. Geothermal, meaning thermal (heat) energy from the geosphere, the Crust (on which we live) and outer Mantle. The Earth has a solid iron Core, surrounded by a turbulent Outer Core of liquefied iron, at a temperature of about 5000 deg. Centigrade, or roughly equivalent to that on the face of the Sun. This heat keeps the Mantle, a plastic mixture of minerals, rolling and writhing, somewhat like the mineral oil in those lava lamps of a few years back. As it rolls and writhes, the Mantle carries heat energy upward towards the Crust where occasionally it breaks through, in volcanoes, in rifts, such as the Mid-Atlantic ridge or extrudes molten rock into crustal formations.

It is unlikely that Man can ever tap into the heat of volcanic activity in a commercial way, although it would provide limitless heat energy, until the volcano died, at least. The mid-oceanic ridges would be harder to tap into, although they have readily available heat transfer medium - water - all around them.

This leaves the intrusive sections of magma or molten rock, and the heat they create. Mostly they can be found around geologically active areas, plate boundaries, volcanic areas etc., and are frequently characterised by hot springs, bubbling mud pits, geysers, like Old Faithful in Yellowstone National Park, a geyser that you could almost set your clock by, the regularity of its eruptions. There are several well known geothermal areas in the world, the major energy producers being USA, Japan, Mexico, Italy and New Zealand. There are more players entering the arena all the time, a table of which is shown in Appendix II.

These magma intrusions can take anything from 5000 to over a million years to cool down and, because the Earth is a living thing, they are constantly being created or renewed. Mankind has used such hot rocks for centuries. Peoples of China, Iceland, Japan, New Zealand, North America and other areas have used hot springs for cooking and bathing. The Romans used geothermal water to treat eye and skin disease and, at Pompeii, to heat buildings. Medieval wars were even fought over lands with hot springs*. Today, as long ago, people still bathe in geothermal waters.

The Japanese tradition of social bathing dates back to ancient Buddhist rituals. Beppu, Japan, has 4,000 hot springs and bathing facilities that attract 12 million tourists a year. Other countries with major spas and hot springs include New Zealand, Mexico and the United States.

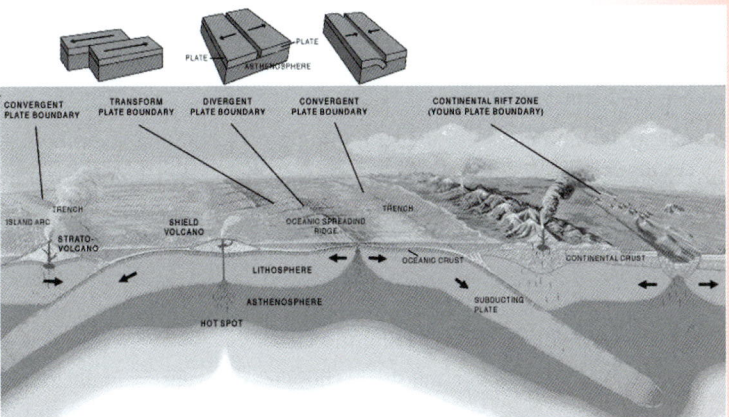

This graphic from the USGS shows how heat from the Mantle reaches the surface by various means, not all of which is of use to man, but much of which could be, such as the hot spots shown rising from the Core. Where they break the surface, we get volcanoes - the Hawaiian island chain is built by one such hot spot - but there are other areas where the hot spots simply increase the temperature of the rocks above and around them. This 'free' heat is available for tapping.

*What a surprise! Sounds familiar, doesn't it?

The heat from these geothermal centres can be used for industrial purposes. It is almost 6,500 kilometres (4,000 miles) from the surface to the centre of the Earth, and the deeper you go, the hotter it gets. The crust is anything from three to 35 miles thick and insulates us from the hot interior. From the surface down through the crust the normal temperature gradient (the increase of temperature with the increase of depth) in the Earth's crust is anywhere between 17-30°C per kilometre of depth (50-87°F per mile), a phenomenon well known to oil drillers, as it creates constant problems with drilling fluids and equipment.

In some regions with high temperature gradients, there are deep subterranean faults and cracks that allow rainwater and snowmelt to seep underground - sometimes for miles*. There the water is heated by the hot rock and circulates back up to the surface, to appear as hot springs, mud pots, geysers, or fumaroles. However, if the ascending hot water meets an impermeable rock layer, the water is trapped underground where it fills the pores and cracks of the surrounding rock, forming a geothermal reservoir. Much hotter than surface hot springs, geothermal reservoirs can reach temperatures of more than 350°C (700°F), and are powerful sources of energy.

We can reach these geothermal reservoirs by drilling wells down into them. After an exploration well confirms a reservoir discovery, production wells are drilled. Hot water and steam shoot up the wells naturally (or are pumped to the surface) at temperatures between 120-370°C (250-700°F) where the heat is used to generate electricity in geothermal power plants. Shallower reservoirs of lower temperature - 21-149°C (70-300°F) - are used directly in health spas, greenhouses, fish farms and industry and in space heating systems for homes, schools and offices.

An inherent problem with geothermal energy however is the fact that steam, much like acid, is a good solvent, dissolving chemicals and minerals from

---

*Did you know that the water that camels (and people) drink in desert oases in the Sahara comes from the mountains to the North of Italy, the Dolomites and the Alps? It is millions of years old!

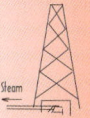

Steam

Water

the rocks it passes through. These are then deposited on equipment and piping as the steam cools, leading to corrosion, blockage and malfunction of equipment. This requires the injection of other chemicals to stabilise those being brought up by the steam, a not inexpensive solution.

Geothermal energy has been treated as the poor brother of the oil and gas industry by most governments and companies, as it uses similar equipment to reach it, it uses similar production methods and like gas, requires almost instant use as it is very difficult to store. More attention will be paid to it, as energy sources dwindle, at least by the 'haves'. The 'have nots' will have to just sit by and watch, maybe pay for geothermally produced electricity from a luckier neighbour, or find their own solutions.

Solutions there are. In fact, it surprises me considerably, after an article I wrote for one of my magazines, in which I described a system which, even at that time was perfectly feasible and which can produce clean, hot water or steam - a recyclable system. We have the technology, such as demonstrated in places like Wytch Farm in England, where horizontal wells can be drilled several kilometres away from the drilling rig, or production well. We have the equipment, with special casing and tools designed to combat high temperature, high pressure production fluids. By drilling and then setting a horizontal casing several kilometres into a hot zone, then completing the well by running a string of tubing inside the casing, a simple circulating system could be set up that recycles purified water, down the long way, [the long way is through the annulus & the

Water circulated down the annulus (the space between the casing and the tubing) returns up the tubing as superheated water in a closed system, reducing the need for chemical treatment of the water.

Hot zone

short way is through the tubing], does not pick up soluble elements on the way (thereby reducing the need for continuous chemical injection) and provides the heat transfer needed to produce electricity.

Going back to an earlier statement, that temperature increases by 17-30°C per kilometre of depth, one well, that was drilled but eventually abandoned by the Russians a few years ago, was more than 12 kilometres deep. So that could give a temperature differential of more than 300°C, more than enough to satisfy anyone's need for increased temperature. It may not be necessary to have a hot zone right under your feet, all it takes is the will, and the effort. Equipment and solutions exist, the question is financial, as with all energy sources. So long as hydrocarbons are relatively cheap and available, geothermal energy is not going to see very much investment. But things will change.

One final aspect of geothermal energy which may have slipped by unnoticed is the fact that it can only be used to produce heat, and electricity.

" Exxon forecasts that by 2030, full 95% of the world's vehicles will still be run on internal combustion energy. "Renewable energy is a complete waste of money,"

Lee Raymond, CEO of ExxonMobil, 2004.

08

NOT EVERYONE
AGREES

As can be seen from the foregoing, Energy, while not exactly in abundance, is available in various forms that can be turned to the use of Mankind. None of it, however, is a permanent solution to the expanding needs of the human race.

Much time and effort, and money, is being put into developing alternative sources of energy. Unfortunately most of it is static, i.e., the energy is suitable only for static uses. If you cast your mind back to earlier in this discourse, as stated before, most energy is used in transportation, whether it is people, or goods – trains and boats and planes, trucks, cars, motor bikes - every conceivable means of using energy extrasomatically, something that we, the human race are past masters at accomplishing. Unfortunately, much of the time, effort and money is being put into energy production systems that cannot even be converted into transport systems. No versatile, high capacity, high power system has yet been discovered that can replace the hydrocarbon energy system we presently have. Cars are being developed which use solar energy recharged batteries. What happens when the sun goes out? None have been built that can haul a 100 tonne road train through the deserts of Australia, or lift a helicopter off the ground, let alone fly a Boeing 747 or an Airbus A380 from one side of the world to the other.

The following chart gives some idea of the collective term 'solar energy' and the range of potential means it covers.

Solar Energy						
Technique	Thermal	Photovoltaic	Photosynthesis			
Primary output	Heat	Electricity	Fixed carbon (biomass)			
Secondary output	Turbines	Hydrogen	Heat	Gaseous fuel	Liquid fuel	Solid fuel
	(Electricity)	(Hydrolysis)		Turbines (Electricity)		

Solar energy panels are notoriously inefficient. They do help in generating hot water, but in converting the sun's energy into electricity in a big way they are basically a non-starter. Research will undoubtedly produce more able, more efficient solar cells (photovoltaic cells) as time goes on, but we really don't have the luxury of time that it is going to require to produce a cell that can convert high double digit energy efficiency. Even so, the US has installed or plans to install by 2010 some 500,000 rooftop solar units, while Japan will have some 4,600 MW of solar generation capacity, a drop in the bucket of Japan's energy consumption.

One of the energy systems that unfortunately comes under the term 'solar energy' is the burning of biomass, the waste product of essentially sugar cane and other annually grown plants. When we talk about biomass, we are referring to a process little removed from Man's first attempts to control fire by utilising vegetative matter, grown through photosynthesis and converting the power of the sun into usable energy. Unfortunately, it also creates almost as much pollution as the original idea, plus it leaves behind a tremendous amount of waste produce requiring disposal. Good for country dwellers, but not much else.

Another energy source from biomass is the biofuel ethanol (made from fermenting grain - the same process used to make sake and other alcoholic beverages…) but it is only suitable to be used in a petrol/ethanol 80/20 blend. However it does have the benefit of maintaining the CO_2 balance instead of releasing historical CO_2 into the atmosphere like hydrocarbon fuels. A report in The Economist in May 2005 states that America is producing 30% more maize-based ethanol a year. Brazil is making it from sugar, while China is building the biggest ethanol plant in the world. Not only that, Germany, the biggest producer of biodiesel, is upping production by 40-50% a year. Sounds great! The problem is solved, right? But look further down the column, total production of biodiesel accounts for only 4.5% of annual consumption. Likewise with ethanol production-maximum output was 4 billion gallons against a fuel demand of 175 billion gallons. Long way to go yet folks. But, like everything else, it is a supplement not a substitute or replacement.

Solar energy also lends itself to another form of energy, and that is wind energy. Wind is actually the transfer of energy between one place and another. When the sun heats up a volume of the atmosphere, it creates a temperature differential between that volume of air and another at a lower temperature. The hotter air, being less dense, (if you remember from earlier, heat causes the average distance between the air molecules to increase thereby reducing the number of molecules in any given volume) begins to rise upwards, creating an area of lower pressure, which in turn, induces denser, colder air to move in to replace the rising column of hot air. That movement creates what we know as wind. It can be gentle, like a cooling summer breeze, or it can be violent, like a hurricane, leaving devastation and destruction in its wake.

Useful air currents, i.e., those that can be entrusted to be stable more or less all year round, such as Ilocos Norte on Luzon Island in the Philippines, can be used, as with the windmills of old, to turn large propellers which in turn, generate electricity from one of several types of conversion machinery. Most of you must, by now, have seen the so-called 'wind farms' with many of these large propellers slowly rotating in the breeze, generating electricity as they go. There are an estimated 50,000 wind turbines in the world (2005), with many more being installed. The critical element in all this is location - location, location, location. Hillary Clinton once said "We have all this empty federal land in Nevada – it should be packed with wind turbines and solar panels."

But, the following story from the BBC News illustrates a point:

Wind farm public inquiry begins

A public inquiry into proposals to site one of the country's largest wind farms in Cumbria begins on Tuesday. Renewable Development Company and West Coast Energy want to erect 27 turbines at Whinash in the Lake District, each almost 400ft high.

The turbines, which would stretch 7km to the Yorkshire Dales, would generate enough power for 46,000 homes. Green groups back the plans, but they are opposed by tourism chiefs who fear a 'dangerous precedent' will be set. The £55m plan would involve turbines being spread between Tebay and Shap overlooking the M6, and would generate 57 megawatts of electricity.

Important step

Developers say they have consulted a number of environmental groups in the run-up to submitting plans. Green lobby groups will highlight local support for the wind farm when the inquiry begins in Penrith. They will also say it is an 'important step in tackling climate change'.

Friends of the Earth and Greenpeace are both backing proposals. The site is near the M6 and lies between the Yorkshire Dales and the Lake District National Park, although it is not a designated national park area.

Tourism 'affected'

Cumbria Tourist Board said it is not against renewable energy. But it claims the building of the wind farm at Whinash could impact on visitor numbers to the area. South Lakeland Friends of the Earth collected more than 700 letters from local people in support of the wind farm and sent these on to the Department of Trade and Industry and to the Public Inquiry Office.

Environmental campaigners will ask the government inspector to disregard claims made by the opponents of the scheme, ruled as 'misleading' by the Advertising Standards Authority, about damage to house prices and the tourist industry in the Lake District. Because of the size of the project the Department for Trade and Industry (DTI) will be responsible for planning consent.

In other words, it's okay in your backyard, but not in mine. Not only that, the English turbines are 400 feet tall, making them a hazard to aircraft.

Also, if you compare figures between the story above and the story below, either British turbines are more efficient, or American turbines are smaller, or their houses are much bigger and they are spendthrift when it comes to power:

UK	vs.	USA
27 turbines	vs.	450 turbines
46,000 homes	vs.	70,000 homes
57 megawatts	vs.	300 megawatts (see below)

Either that or someone's math is sadly lacking. Do they know what they are talking about?

'Good positions for wind farms are also often sites of great natural beauty.... The answer may be to site wind turbines away from towns and cities, for example in shallow coastal waters...' says Ben Russell of the UK Science Museum in 'A Question of Energy: Fossil Fuels, Renewable Resources and Wind'.

The biggest wind farm in the world is being built in Oregon, USA and consists of 450 wind turbines - four hundred and fifty - to produce 300 MW of electricity, enough to power 70,000 homes. California, with some 17,000 wind turbines installed can power a city of 300,000 at peak output.

Some quick maths will show that, if we assume, let's say three people per household average, that's 17,000 turbines for 100,000 households, so we'd need 1,020,000,000 - 1.02 billion - wind turbines to keep 6 billion people happy. Mind you, that's running at peak output. Another massive project is the £1.5bn, 1 GW London Array wind farm that could see some 270 turbines spread over 152 sq miles (245 sq km) in the Greater Thames Estuary in Southern England.

The problem however with wind farms is that the wind doesn't always blow when you want it to so you have to build sufficient turbines to meet minimum demand; you can't store electricity, so you have to feed excess into the national grid; you cannot always predict wind speed, so elaborate speed controls are necessary on the windmills to prevent them from being blown apart by hurricane-force winds; but lastly, and more importantly, they can easily get damaged. Because of their huge diameter (up to 40 metres across), they are prone to self destruct – the tips of the 'wings' can reach the speed of sound at quite slow revolutions per minute (rpm), setting up vibrations that can literally shake the huge propeller apart.

Similar 'propellers', using water power, have successfully been used in tidal areas, especially fiords, where the tidal movement is massive and rapid, to generate electricity from the coming and going of the tide, all they have to do is keep shipping away from them!

Another recent invention is the wave generator. The earth is mostly covered with water (some 70% of it!) which picks up energy, sometimes lots of it, from the winds that circle the planet. Anyone ever having seen surfers risking their lives on these huge waves can appreciate the kind of energy that winds can impart to a moving body of water. Such wave power is of course too violent and would destroy anything built to try and tap its energy, but there are areas of the oceans, estuaries and such like that have a prevailing wave form that can be utilised to produce power, in this case electric power from wave movement. By utilising specially designed floats that bob up and down as waves pass them, these static generators can be used to generate electricity.

Again, by utilising the temperature difference in sea water, scientists have also devised a generator that used the solar energy that heats the water to create a cycle that drives a turbine, thereby generating electricity.

Tides move large masses of water back and forwards and generators of various designs and efficiencies have been designed and installed to try and benefit from such movement, such as the Oscillating Water Column, where the tide pushes water up a column, which in turn forces displaced air around a turbine. As the tide recedes, it works in reverse, sucking air back into the column, turning the turbine as it goes. The use of tides, caused by the forces of the Earth, Moon and Sun as they rotate around each other, has been known to man since the turn of the first millennium, i.e., the 10th or 11th century.

There are many ingenious ways of producing electricity from a variety of sources by a variety of methods. But in all these installations, the end result is the same, the generation of electricity to support a national grid demand. None of it will get you off the ground or to the other side of the world. And this is where the Earth is going to face its greatest dilemma.

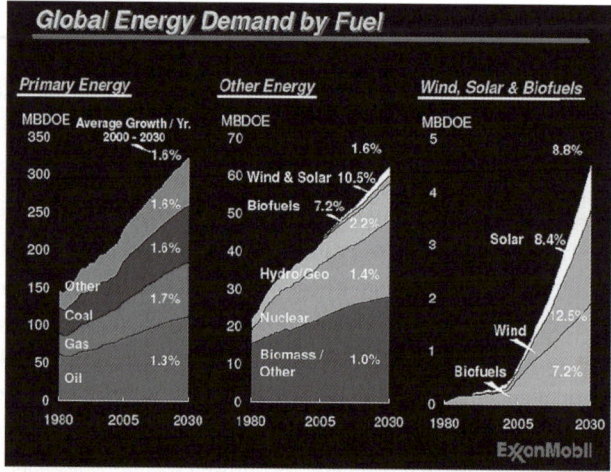

A quick glance at the values (MBDOE) of the columns will let you see that hydrocarbons lead the field by a wide margin. No other energy sources will ever match the output of oil & gas – i.e., it's a one way street, downhill.

■ Where we go from here

...Where it is Leading

So where do we go from here? Anyone who knows the answer to that question can name his own ticket. One of the requirements for sure is a more concerted effort by all of Mankind to share what little hydrocarbon energy resources remain to us.

You cannot look at this situation in isolation, we have to look at it from a global perspective. We are after all, talking about the continuation of the human race. Whatever we do to this planet - pump billions of tons of CO_2 and other gases into its atmosphere; heat it or freeze it; strip it of vegetation; poison its waters; deplete its chemicals resources - it will survive. The planet's ecological and environmental systems will tend towards a state of equilibrium - in a decade, a century or a millennium - however long it takes to reach a stable condition. The question is, can Mankind live in the conditions that will exist?

Let us analyse the situation and look at the possible pathways into the future remaining to Mankind. The theme of this book has been - all along - the Energy Trail; how we got started on the trail, how we developed along the trail and where the trail is going to lead us to in the end. There is really only one direction for Mankind to go in now, if you think of the consequences of our own greed, and that is a reversal of the trail we followed in getting here.

If you remember from earlier in the book, we looked at the global population and how it had grown. More especially, we looked at how it had grown with the advent of usable energy on a vast scale… Previous to that, the world's population was based on what a limited agricultural area could support. With the discovery and utilisation of hydrocarbon energy to provide transportation, food, goods and energy could be moved to people centres, massive townships and cities where the population mass manufactured products for sale, while someone else brought them food, light etc., for their everyday living comforts.

The energy requirements of a static situation can be met by a variety of means, mostly electrical, provided by one generation method or another.

The electricity allowed heating and cooling of densely populated areas. It allowed lighting of homes and factories, it allowed massively high rise buildings to be built by the introduction of escalators and elevators. It provided the means to build highly complex people moving systems such as tramways, underground trains, intercity trains, etc.

But take away the electricity. Now take another look at the situation. Look at what happened to north eastern United States and Canada when a blackout hit a few years back - in 2003 - absolute chaos. Look at your own local situation. Have you ever had a power failure? Blown a fuse and had to sit in the dark, with no air-conditioning, no heating? Imagine that on a permanent basis. What would it do to your lifestyle? Never mind that, what would it do to your office, your factory, your warehouse? Everything depends on energy, power, electricity - the doors won't open, they're electronically controlled. The machines don't run. They're electrical. You can't even get to work, eight kilometres away, because the trams and trains don't run - they're electrical.

Mind you, there's not much point in going to work anyway, because the products, even if you could manufacture them, would need to be delivered to a shipping centre, and they can't distribute them because all the transportation is out - diesel engines need diesel, gasoline engines need gasoline, electric trucks need electrical chargers… The list goes on.

What do we do when we run out of energy resources? Look back at the list of possible energy resources - electricity, provided by some generation method, usually indirectly from inefficient heat converters which burn a fossil fuel, oil, gas or coal, or bagass, or nuclear, converting steam pressure into rotation through steam turbines - all of which have a very limited lifetime, except perhaps bagass.

The principal word here is electricity. We use our energy resources to produce electricity which by its very nature is only of use on a limited basis. We cannot fly aeroplanes on electricity, but only a small proportion of global merchandise moves by air anyway. Most of it goes by sea. We

Our problems are going to be compounded by the growing demand of those who do not yet have the basic necessities of life, clean water, lighting, sanitation, etc. Do they not also deserve to be brought into the 21st century? China's rising demand for energy is driven by the liberalisation of its policies and the natural entrepreneurship of its people. India will follow suit, and it has a much bigger entrepreneurial base. What of all the other 'Third World' countries, in Africa, the Middle East, Asia, South America, don't all the people of these countries also deserve a share of the diminishing energy resources of our planet?

can drive ships with electricity if, and this brings us back to the same position as the land-based power providers, we have the fossil or nuclear fuel to create the electricity.

Our problems are going to be compounded by the growing demand of those who do not yet have the basic necessities of life, clean water, lighting, sanitation, etc. Do they not also deserve to be brought into the 21st century? China's rising demand for energy is driven by the liberalisation of its policies and the natural entrepreneurship of its people. India will follow suit, and it has a much bigger entrepreneurial base. What of all the other 'Third World' countries, in Africa, the Middle East, Asia, South America, don't all the people of these countries also deserve a share of the diminishing energy resources of our planet?

Right now, in 2007, we are squabbling about oil production, the price, how much the world is consuming (about 86 million barrels a day). As the years go on and these squabbles become fist fights, the fist fights will become more serious and we can expect to see war break out over possession of remaining resources. This is not a prophecy, this is a fact. Whatever grounds were given for the Allied invasion of Iraq, Saddam Hussein's removal was probably of low priority compared to the securing of Iraq's oil and gas reserves for Western interests.

In years to come, in years that can be counted on two hands at the most, we will see more and more aggression, in more and more locations. It is reaching a stage where Darwin's theory of Survival of the Fittest takes on a whole new meaning and will devolve eventually into one word - Survival. It probably won't happen in your lifetime, definitely not in mine, but my children and their children will have to learn to live in not so much a Brave New World, as a completely alien world, where survival will become of prime importance.

One thing we all seem to have forgotten, or at least have put aside, is the fact that this planet doesn't owe us a living. We evolved on this planet along with many other species, we were fortunate (?) in developing the ability to make use of our environment and use it we did, to the point that we are the most destructive creature ever to roam this planet. Ants are more numerous, and (probably more intelligent), dinosaurs were more in tune with their environment (they lasted hundreds of millions of years, we've been around, what? A couple of million at the very most?). Almost any creature on the face of this planet that you care to name has a better rapport with it than we have.

We have been enormously successful. Or have we? What is the measure of success? Is it all the achievements of we humans, or is it the evolution and development of the human race? Is it longevity, or is it control? What is the point in being a highly developed, highly successful species if we only survive for a few million years? The Earth will survive until the point a decaying Sun consumes it in a fiery hell, many billion of years from now. How many of these billions of years will we share with this planet? None. Not even one. I remember, quite some time ago, seeing a 24 hour clock face supposed to depict the life of the Earth. While the 24 hours represented the Earth's past - Homo Sapiens arrives on the scene somewhere around a tenth of a second to midnight. The details are irrelevant, what it serves to show is how minute a part we have played in the Earth history. But the damage we have wrought outweighs anything that has ever happened to the planet in its long history, whether meteorite strike, volcanic eruption, earthquake or other natural phenomenon. We have stripped the planet of its vegetation, we have disgorged billions of tons of carbon dioxide and other historical gases back into its atmosphere, we have changed its climate - the list is endless - and we think we're successful?

The human race, as we know it, will be lucky to survive to the end of this millennium. Assuming no marvellous breakthrough in energy provision occurs we will follow a slow devolution back down the road from whence we came. As we use up the available finite resources, the first thing that will affect us is the inability to get around as easily as we used to.

With no fossil fuels, transport will become a serious problem. I'm not talking about getting down to the village shop, I'm talking about getting from one part of the globe to another. Aeroplanes will stop flying because no power unit has been developed, nor will be developed in time, to fly these behemoths around the world without hydrocarbon fuels. This leaves shipping. An article in The Economist stated that 'If Trade is the lifeblood of the world economy, then ships that perform the mundane task of transporting goods and raw materials from where they are produced to where they are wanted are the red corpuscles. In 2004 the world's fleets carried around 90% of total global exports worth US$ 8.9 trillion, largely unnoticed.'[13]

While ships will survive so long as a nuclear power source is available, in the long run, these too will cease to be viable - who needs a container ship when there is nothing to ship?

Yes, nothing to ship. What is shipped around the world? Food. Machinery. Coal, oil & gas. Grains. TVs, stereos, cars - we're living in a market oriented world. When there is no market, there is no need for produce, no need for trade and industry. Remember, we are going to be living in a world where energy is so scarce that most of it is going to come from our own backs. No electricity means no big cities. People will move back to the country to try to support themselves and their families. People moved to the towns, the big cities to find work, to trade, to buy produce, to offer their services. When there is no city, no work, no trade there will be no point in congregating in large groups. Survival becomes more important.

One of the great discoveries of Man was fire. It allowed him to venture much farther north on the landscape towards the North Pole than his predecessors. He found new ways of surviving. In the arid desert regions of the equatorial belt of the Earth, food was difficult to come by. Populations developed around the areas with plentiful food and ground suitable to grow produce. By venturing northwards, Mankind opened up a much larger area for development.

But today, without energy at the flick of a switch, how long will people survive in these frigid regions? The Inuit survive by living in harmony with nature and the surrounding land. How many others could survive under such conditions? Where it is too cold to survive, people will migrate towards the equator, just the reverse of what their ancestors did a million years ago. People live in -30°C because there is energy to keep them alive. Without that 'potted' energy, who will live in conditions such as that? As the sources of energy disappear, as society breaks down, people will once again become more isolated, more centric in their living habits, self first.

Unless a renewable, efficient source of transportable energy is found and found within a few hundred years, society is going to break down and we will revert to the same conditions as existed thousands of years ago when Man first ventured out onto this planet. Only then, instead of the wanderlust to travel, to go places, to explore that drove the great adventurers of old, people will live in enclaves where trust doesn't exist, not even amongst the members of the individual enclave. A movie from 1995, *Waterworld* starring Kevin Costner, is not really so farfetched if you understand the theme of the movie rather than marvelling at the filming technique. The *Mad Max* movies starring Mel Gibson are also understandable if you consider the theme. Whether the human race reaches these conditions by nuclear war (still possible) or simply by running out of suitable energy resources, these movies might well be prophetic.

Anyone having reached this far in this book will by now I hope, understand what I have tried to do in the telling of this story. We are a product of this planet. We have no more right to be here than any other species. We have evolved through our own or our forebears ingenuity to a point where we are the most widely spread, most dangerous creature on this planet. We got here through the development and drive of our own capabilities. We evolved, in a matter of less than two thousand years from a race of a few million souls, to a population of more than six billion people. Simple arithmetic will show that to feed a few hundred people, most of whom worked in the field tending to crops, is quite possible. But increase that to six billion people, even if the majority of them worked in the field, it would be impractical to house and feed such an enormous population.

Feeding such a huge population was achievable through the mechanisation of farming. In America, less than two percent of the population feeds not only the peoples of America, but a large proportion of the rest of the world. Take away the mechanisation. Take away the grain harvester that covers hundreds of miles every day, reaping millions of acres of wheat, barley, oats, maize, etc., to feed the masses. Take away all the mechanised machinery. Return to the labour of one's back, return to the farming techniques used before energy was so freely available. How many people could be fed? One billion? Five hundred million?

Spread around the equable regions of this planet, and using techniques not available to our forefathers, the human race could probably support somewhere around that number of people. A far cry from the six and a half billion and counting that we have today. Even if we could somehow reach a consensus to conserve all the remaining portable (i.e., fossil) energy on this planet, all that we would be able to do would be to slow the decline in living conditions.

One of the problems facing us right now is the fact that we are not only stripping the planet of the hydrocarbon energy still remaining, we are accelerating the use of this energy to the point that we are simply hastening towards the point of no return. There are numerous publications[14] and websites[15] that have been sounding warning bells, some for many years and yet, few members of the public seem to care, or even be aware that such a crisis is looming. Like the builders, the captain and officers of the Titanic who were so sure their ship was invincible, we are sailing into iceberg infested waters with no knowledge of how deep or far reaching are the problems we will face very shortly.

"Energy has always been the basis of cultural complexity and it always will be. … the past clarifies potential paths to the future. One often-discussed path is cultural and economic simplicity and lower energy costs. This could come about through the 'crash' that many fear - a genuine collapse over a period of one or two generations, with much violence, starvation and loss of population. The alternative is the 'soft landing' that many people hope for - a voluntary change to solar energy and green fuels, energy-conserving technologies and less overall consumption. This is a utopian alternative that, as suggested above, will come about only if severe, prolonged hardship in industrial nations makes it attractive and if economic growth and consumerism can be removed from the realm of ideology."

- Joseph A. Tainter[11]

09

IT STILL EXISTS,
THE QUESTION IS COST

Are we really running out of hydrocarbon energy? A straight answer would be 'No. We will never ' run out of ' hydrocarbon energy. What we will run out of are people willing to pay the cost of recovery. Eventually it will become so horrendously expensive to retrieve the remaining oil or gas or coal, we will not be able to grow trees fast enough to replace the ones being felled for fuel, that to all intents and purposes, it will be used up, no longer available. A report from the World Energy Outlook said that '... fully 84% of our energy demands are being met by hydrocarbons and will remain so until around 2030. Oil demand will actually increase – to reach 116 MMb/d – or 37% up on consumption in 2006. Coal use will increase by 73% - seventy three percent – between 2005 and 2040! It will take some US$22 Trillion in investments to supply the infrastructure to meet this projected global demand – provided the supply can be found...!' In other words, present trends dictate these figures with no guarantee that the resources will be found to meet such a demand!

The last remaining source of hydrocarbon energy not yet tapped by Mankind, is a substance known as a clathrate, or gas hydrate. Under specific circumstances, water will freeze in a cubic state rather than the hexagonal state more familiar to us in the beautiful forms of snowflakes. These crystalline solids, which look like slushy ice, occur when water molecules form a *cage-like structure* around a few gas molecules. The most common are methane, ethane, propane, butane, isobutene, nitrogen, carbon dioxide and hydrogen sulphide. Methane is the most abundant in natural hydrates. Under normal conditions, one cubic metre of hydrate can contain up to 164 m^3 of methane.

Hydrates of this nature are not from the original oil or gas depositions that formed the oil and gas fields we are emptying so rapidly, but rather from bacterial action on organic matter that has sunk to the ocean floor. They also need sufficient pressure to maintain them while a temperature close to freezing is also required. Areas that satisfy these needs are continental margins and elevated seas at low temperature. As the world is made of continents, and the largest land masses are in the polar regions, it would seem that there should be almost limitless quantities of gas - and there are, according to most estimates, measured in gigatons, or sufficient to last several hundred years at present consumption levels. It is also a renewable resource.

The hydrates exist from a few centimetres beneath the seabed, to several hundred metres on average, the deepest being around 2000 metres. The question that arises however, is how do we harvest such a gas by a controlled, economically acceptable means? It is vastly different to having a gas reservoir, where the gas drives itself out of the well because of the pressure. Nor is it like oil, where you can attach a pump and suck the oil up a tube. It is most akin to nodule mining, if anyone remembers the big news stories of years ago about mining manganese nodules from the sea floor, an ideal source of revenue, but excruciatingly expensive to mine. 'Mining' hydrates will hold out similarly tantalising prospects - and costs.

One of the problems facing the 'would be' prospector is the highly sensitive stability of the hydrates. They exist in a zone of stability where temperature and pressure are suitable for their existence. Change either of these and you have a problem. Increase the temperature and the ice will melt, leaving a gaseous ooze behind that will slowly lose the gas into the sea. Reduce the pressure, and there is a high risk of rapid gas release. As the pressure reduces, the gas molecules expand, breaking the weak links that hold the ice cubes together, releasing gas bubbles into the ocean. This released gas reduces the water density, decreasing the pressure even more and you end up with a runaway effect, similar to a runaway nuclear reaction, or cascade. Investigations of the coastal regions of the USA have indicated that large scale slumping, or landslides have occurred on the offshore shelf area, thought to be due to hydrate release and subsequent weakening of the undersoil caused by pressure reduction. Such pressure reduction could be caused by an advancing Ice Age sequestering water into ice shelves, thereby reducing water pressure on the hydrates.

Methane can stay in the atmosphere for around ten years (it slowly oxidises to CO_2) but in the interim period it acts just like any other greenhouse gas, bringing about a global temperature rise, except that methane is ten times as effective as CO_2 in its greenhouse effect. So, harvesting hydrates could in fact hasten the end of the Earth as we know it instead of providing an alternative gas resource for our diminishing hydrocarbon reserves.

Earlier on, I talked about nuclear energy and the fact that there was only sufficient Uranium to last another two hundred years or so. One nuclear energy source which I have not so far mentioned may prove to be the saving grace of the human race in the long run - if we can learn to master it - and that is nuclear fusion.

If you remember, the nuclear energy we talked about before was the result of the breakdown of individual radioactive atoms as they tried to find a stable state. This is termed fission, or breaking apart, and the materials are classified as fissionable materials or are said to be fissile, from the Latin fissus, to cleave or split. It is a destructive process and eventually the material will run out of atoms to split and the process will come to an end.

A more enduring process is that of fusion, the process that has kept our Sun functioning for at least 5 billion years and the Universe as a whole for a few billion years before that. Fusion simply means to join together, in a permanent fashion. You could hold two tennis balls together, but that wouldn't be fusion, not unless you could get all the atoms of each ball to intermingle with each other such that you couldn't separate one ball from the other any longer. That would be fusion.

It happens on a grand scale every second of our lives in the fiery body we call the Sun. The Sun is an agglomeration of gases and particles, held together by the enormous gravity of the agglomeration. The majority of particles in the Sun are Hydrogen and Helium nuclei, although there are other elemental particles (between them, hydrogen and helium account for 99.9%). As these nuclei get crushed together under the enormous gravitational attraction, they collide into and bounce off of each other and once in a while, probably in a head on collision, the two nuclei fuse together and stay together.

So now, if we just think for a moment, a hydrogen nucleus is what? It's a proton, right? The nucleus of an atom is the central part, the core of the atom without the cloud of electrons. The pressure and heat inside the Sun is so great that even the heavier atoms are completely stripped of their electrons and the gas becomes what is called plasma. This is a completely different phase of material existence. If you remember, we talked about the progression of heat content, from solid, to liquid to gas? Well, plasma goes beyond even the gas stage and becomes a completely new entity, a 'soup' of nuclei, electrons, neutral particles and such, existing at a temperature measured in millions of degrees. Everything, EVERYTHING would have become a gas long before it reached the core of the Sun, which is about 15 million degrees Centigrade, where all the action takes place. The core of the Sun is actually a huge fusion reactor, and the Sun is really a balance of gravitational force, trying to collapse the star, and a runaway nuclear reaction trying to blow it apart! Something to think about as you lie in bed of nights!

It is relatively safe however - it's had at least 4.5 billion (4,500,000,000) years to blow itself apart but it hasn't done so yet. Scientists reckon it will eventually, some 4.5 to 5 billion years from now.

Even the surface of the Sun, although a balmy 6,000 degrees Centigrade plays a part in our lives as this is where we get light and heat from. The darker spots on the surface of the Sun, so called 'sunspots' are a lot cooler, around 4,000 degrees Centigrade! The light and heat radiated from its surface came from the Sun's core and took a million years or more to reach the surface. In the process of producing this light and heat, the Sun 'converts' 700 million tons a second of hydrogen into helium, in the process releasing five million tons of energy.

In a simplified description of the fusion process known as the P-P chain, the dominant process in most stars, pairs of protons fuse, forming deuterons*. Each deuteron fuses with an additional proton to form helium-3. The two helium-3 nuclei then fuse to create beryllium-6, which is unstable and disintegrates into two protons plus a helium-4 (alpha particle). In addition, the process releases two positrons, and gamma rays. The positrons annihilate quickly with electrons in the plasma, releasing additional energy in the form of gamma rays.

Can you see where this would lead us? Eventually, there will be more helium nuclei than hydrogen nuclei and helium nuclei will be more likely to bump into other helium nuclei than hydrogen nuclei. Although hydrogen still accounts for 91% of the Sun's volume, a time will come when this helium to helium collisions will become more prevalent and the Sun will enter what is referred to as the Helium cycle, or Helium phase. This phase will herald the end of the Sun as we know it, but then, we probably will be long gone before that happens.

The purpose of this description is to give the reader some idea of how enormous a task it is to create nuclear fusion here on Earth, a process that happens in the core of a sun. In comparison, nuclear fission is a bit like falling off a bike. All you need is the right conditions.

While many attempts have been made - many scientists have dedicated their lives to it, unimaginable sums of money have been spent on it, enormously complex facilities have been built to handle it - nobody has as yet succeeded in containing and propagating a reaction in the million degrees plus conditions necessary to produce a sustainable fusion reaction. One of the (many) problems is producing the pressure needed (the Sun's core is something like 340 billion times atmospheric pressure), not to mention the ingredients (a plasma rich in hydrogen nuclei) and the temperature (a million or more degrees). The containment seems to be

*There are actually three forms of hydrogen, a single proton, a proton plus a neutron, called deuterium, and a proton with two neutrons, called tritium. Since unstable nuclei can only find stability downwards, a combination of two particles is referred to as deuterons, or deuterium nuclei, not helium.

best accomplished through a magnetic bottle*, a highly complex piece of machinery that holds the plasma (remember, charged particles) in a magnetic field while it is bombarded with high energy protons fired from a machine called a cyclotron or cyclic accelerator - a bit like shooting a .45 magnum bullet into a jostling crowd of similar sized bullets moving around at hundreds of miles and hour, and hoping to get a direct hit.

It's not impossible. Scientists at the RHIC in New York[†] have succeeded in directing beams of gold nuclei, travelling at close to the speed of light, towards each other and smashed them together. The intense heat of the collision breaks down the nuclei into the most basic building blocks, forming a ball of plasma about 300 million times hotter than the surface of the sun. The fireball lasted for only 10^{-23} seconds however. Long enough to get measurements off but useless as an energy source - nor can we afford to use gold as an energy source. Not only that, the energy required to accelerate the ions to the speeds necessary to create these collisions of annihilation are orders of magnitude greater than the energy produced (see footnote).

Going back to our original idea of a fusion reactor, the fuel is both abundant, and scarce. To avoid the need for four protons colliding all at the same time with sufficient energy to fuse, scientists use deuterium, an isotope of hydrogen (remember, same proton count, but extra neutrons, to give a different weight) and tritium, another isotope of hydrogen. The trouble is, while deuterium can be extracted from seawater (this is the heavy water used in heavy water reactors), tritium with a half life of only 10 years, is pretty hard to come by. It can be produced by bombarding lithium (see the Periodic Table Appendix IV) with neutrons. The only problem with that is, it requires a fission reactor to produce the neutrons to bombard the lithium! Sure does become complicated.

---

*The most efficient appears to be the Tokomak design. However, any form of magnetic containment requires enormous quantities of energy to produce the magnetic fields and accelerate the 'bullets'. Another Catch 22 situation - the whole idea being to produce energy, not consume it!

[†]Relativistic Heavy Ion Collider (RHIC) at Brookhaven National Laboratory in Upton, New York, US

And this is where its ability, to lift Mankind out of the downward energy spiral he will find himself in, will fail. Unless fusion energy can be contained, mastered and commercialised, it will become just another white elephant along with so many other brilliant but impractical ideas. Even so, as I write, agreements are being reached to build a multi-billion dollar fusion reactor in the south of France, at Cadarache, near Marseille. Conservative estimates put, (if successful !!) first production at about 50 years. Cutting it a bit close don't you think?

"It has often been said that, if the human species fails to make a go of it here on the Earth, some other species will take over the running. In the sense of developing intelligence this is not correct. We have or soon will have, exhausted the necessary physical prerequisites so far as this planet is concerned. With coal gone, oil gone, high-grade metallic ores gone, no species however competent can make the long climb from primitive conditions to high-level technology. This is a one-shot affair. If we fail, this planetary system fails so far as intelligence is concerned. The same will be true of other planetary systems. On each of them there will be one chance, and one chance only."

Fred Hoyle[19]

10

TIME TO LISTEN, NOT TALK

One of the most respected news journals in the world is *The Economist* from the UK. On the 30th anniversary of the Arab oil embargo of 17 October 1973, it ran a telling article on our continued dependence on oil, especially Arab oil. It is an article well worth reading. The leader, 'The End of the Oil Age', to the article, started off by overstating the obvious, that *'the Stone Age did not end for lack of stone, nor will the Oil Age end for lack of oil.'*

It was a statement made by Sheikh Zaki Yamani, the Oil Minister for Saudi Arabia during the oil embargo by Middle East OPEC members of principally the United States and its European & Japanese allies for their support of Israel during the Yom Kippur War which had begun on 6 October of that year. Definitely not the first time that oil had been used as a threat and probably not the last either. However, the results of that embargo, along with the increases in crude oil prices demanded by OPEC are well known today, for they profoundly affected not only the history of the oil & gas industry but the history of the world itself. A second oil price shock hit the world several years later when the revolution in Iran caused another wave of panic in the oil markets of the world.

After the initial panic in 1973 (remember, Saudi was the principal source of US oil), many changes took place throughout the world. Where once the oil & gas industries were the domain of the large 'Seven Sisters' companies, suddenly governments became involved and national oil companies sprang up all over the world, Companies like Petronas and Petrovietnam, both started less than two years after the embargo, were governments' 'insurance' against ever being held up at the point of an Arab gasoline bowser again! Petronas especially, is now a major in its own right alongside companies like Exxon and Shell. As Formula One sponsors for the BMW Sauber F1 team the company has become a household name throughout the world!

Another effect the embargo and rise in prices had on the world was it created a more conservative approach to this valuable source of energy. Did you know that the 55 mph speed limit in the States is courtesy of Sheik Yamani? When prices began to rise, the US government enforced a limit of 55 mph to try to conserve fuel. It not only did that, but road death rates fell dramatically as well. At 100 mph, although wide awake (you better be at that speed!), if you were involved in an accident, you probably killed all the inhabitants of the car and anyone else involved in the collision. At 55 mph you can fall asleep at the wheel and run off the road into a roadside cafe, killing all the people there but you yourself will probably survive. The oil price is continuing to rise, but not the speed limit.

One of the long-term effects of the embargo was it created an economic recession throughout the world. Inflation passed and remained above ten percent and unemployment was at a record high. The era of economic growth which had been in effect since World War II had been brought to an end by a single act – the closing of an oil tap. Today, we appreciate the benefits of the steps taken during that period. The equipment that survived the crisis is three or more times as efficient as it was pre-1973. Imagine the fuel consumption today if nothing had been done to improve efficiency! It forced the Americans in particular to take a closer look at some of their gas-guzzling behemoths, the five- and seven-litre-engined cars that now were cruising the roads at 55 mph consumed (wasted) a tremendous amount of fuel. The people of America slowly began to appreciate the smaller European and Japanese car models. The world seemed to learn its lesson, it began to curb its wasteful ways.

Small start-up companies began searching for oil – the oil majors were looking farther afield, leaving many opportunities to the smaller firms. Many of them found oil, in which case, they were inevitably bought out by one or other of the majors, depending on their profitability. Technological advances made exploring and recovering oil possible in some very inaccessible places, such as deep sea. Initially, floating drill vessels had to be anchored to the sea bed to keep them in place, limiting exploration depth to the few hundred feet a spread of chains or hawsers could hold a vessel in place. But the advent of dynamic positioning, thrusters front and aft of the ship that could hold it in position through GPS indefinitely in any depth of water, meant the ships could do away with their anchor chains. And of course drill in much deeper water. Drillships today can easily operate in more than 10,000ft of water, drilling 40,000 ft of hole (that's roughly 10 miles of pipe!) Which makes oil discovery today very expensive.

Remember, the Ghawar field in Saudi Arabia is still producing oil at a few pennies a barrel. At 280km by 30km it was the biggest oilfield in the world. But with the proving up of the Bakken field in Canada and North America, a field that is more than 1000km wide, it may lose that title. Bakken, discovered in the early 50s may be a tough one to crack though as it is an oil shale region and difficult to produce. Shell gave up on it years ago and went to the Gulf of Mexico, but new technology may change the production capabilities.

So we are still finding oil. As was said earlier, we will never run out of oil, we will only run out of people willing to pay for it. A deeper look into the history of OPEC will show that OPEC was initially formed to protect the interests of the oil producing countries, which were unilaterally being squeezed by Western powers, notably the United States and Great Britain, to reduce oil prices, given the post-war glut of oil available.* Wanting to improve the standard of living in their own countries, the oil producers tried in vain to gain some control over the price of production which was basically in the hands of the oil majors. A rather weak cartel formed in an effort to increase revenue from oil production but to little effect. Remember, they were dealing with the 'Seven Sisters' who controlled world oil production, what price a few Middle East producing countries?

---

*The Concise Encyclopaedia of Economics http://www.econlib.org/library/Enc/OPEC.html

All the investment, or the majority of it, came from the Seven Sisters. Then came 1973. The OPEC of the mid-60s suddenly developed muscle and began applying it. They had been talking for so long, but nobody was listening. Now everybody sat up and listened. They had to! Their cheap source of hydrocarbon energy, on which the world was, and still is, almost completely dependent, had been strangled. A whole generation of people grew up watchful and wary of this producers' cartel as more and more oil producers joined OPEC. OPEC was treated like a pariah, the Arabs, and other oil producers were branded as greedy, money hungry cartel members, out to line their own pockets from the exorbitant oil revenues they were earning at our expense. Oil companies, at government insistence, spend billions of dollars looking for and producing oil and gas from all over the world, to try and reduce our dependence on OPEC, and especially Middle East resources.

However, from another article (*The Economist Oct 23rd 2003, OPEC - Still holding customers over a barrel*)* it is obvious that the world has not gotten over its love affair with Arab oil. The fact that OPEC held around 38-40% of global reserves at that time meant that the world could not turn their collective backs on the Arabs, even though they knew they had been held over a barrel once before and are quite likely to be again in the future. As the Economist article says, two-thirds of proven oil reserves were held by Saudi and four of its neighbours. Converting that quantity of oil into petrodollars, according to an American government department assessment, a staggering US$7 trillion in cash and other funds has changed hands over the last thirty years since the embargo.

So what has changed? In retrospect, nothing really. Oil is like a drug. We can't live without it and like any addict we'll do almost anything to get our hands on it. China's massive demands for oil and gas, as well as Japan – in fact most of Asia - makes the region highly dependent on OPEC as a supplier. Even Indonesia, a member of OPEC, is a net importer,

* http://www.economist.com/displaystory.cfm?story id=2155405

simply because the country cannot meet its own energy demands. The United States is still the largest imported of Middle East crude, consuming roughly 25% of global production. A quick visit to the OPEC website (www.opec.org) will highlight some interesting facts not least of which is that in 2006 OPEC claimed to hold 900 billion barrels of remaining oil. Now forget about the hype and the re-evaluation going on all over the world, by analysts, by governments, by oil & gas think-tanks. Assuming we can hold consumption to 90 MMb a day, (we are already being told it will go to 125 MMb a day before 2025) that gives 10,000 days of oil left in the OPEC tank. That works out to around 27 years - and, since the figures come from 2006 - that means OPEC has enough oil to last until 2033.

"It is part of OPEC's commitment to support market stability, a pledge that goes back to its inaugural meeting in Baghdad in September 1960. This commitment is enshrined in the OPEC Statute, adopted in January 1961, and remains a key guiding objective of the Organization. The Outlook further advances this process and provides a platform from which to review, analyse and evaluate scenarios as to how the oil scene may develop. This should help create a forum for discussion that will hopefully aid dialogue and co-operation amongst all stakeholders, something on which OPEC places much credence.'*

Of course, there is always the possibility, as has happened before, that OPEC could reassess its holdings and all of a sudden 'find' a major increase in oil estimates and add a few hundred billion barrels to its reserves. For an interesting discussion on this check out *'Peak Oil Debunked'* on the web.

There's an old saying that you can't listen with your mouth open. Maybe more than a few people need to stop talking and instead listen to what has been said for so long now by those who are in a position to know the real facts, have information to share, who don't have an agenda to brainwash the general public so that they can remain in power. Money

---

*The Concise Encyclopaedia of Economics http://www.econlib.org/library/Enc/OPEC.html

being spent on maintaining the *Oil King* could be better spent in research and development into alternative energy sources. The Oil King *will* die, and along with it, a way of life.

Unquestionably, we are rapidly running out of resources as well as alternatives. Appendix III gives a breakdown of the available energy sources and what they are doing to the planet we call home.

As with energy conversion, you can't get something for nothing and the process of converting one form of energy into another always leaves a residual effect. In the table, across the top (X axis) there are energy sources, down the side (Y axis) are effects. See how many each source leaves behind.

And this is the essence of what I have been trying to get across in this book. We are using up our hydrocarbon reserves as though they were free. Nothing is free, as we are now beginning to realise, as we witness the changing conditions of our atmosphere, of our planet. It is going to hit us hard, a lot harder than anyone can imagine. Especially if we do not begin to adopt and adapt to the changes that will occur.

This is not intended to be a scare tactics book, trying to get people to change their habits by scaring them. Nor is this one of those types of book of the Eric von Däniken genre, trying to use quasi-scientific facts and outrageous postulation to prove an untenable point. I have not written this book to provide a step by step guide to our future either. What I have done is laid out the facts without any trimmings leaving it up to the reader to form his or her own conclusions. Everything I have written can be crosschecked and verified.

A lot will depend on what we as human beings do, what our governments do, and what the planet itself does. But the final results are anything but inconclusive. They will happen. When, is still open to speculation, but it is in foreseeable time in the not too distant future. A millennium from now will see whether the human race will survive. I fervently hope and believe that the human race has the intelligence to find a way out of this. Our lifestyles will change dramatically. Our numbers will diminish substantially as well. The Sun's outpouring of energy is not going to stop any time soon, we still will have that source.

The planet will begin to assert its own balancing forces, forces that we will be powerless to stop. The outpouring of pollution and heat from industry to support the lifestyles of billions of people is not the same as a few million people living off the land. The planet is slow to react, but react it will. We exploded onto this planet in less than two centuries, with all our accompanying trash and pollution. Two hundred years is a blink of an eye for a planet like the Earth. If the ice caps melt because of the greenhouse effect, most major cities of the world will be under water. If it goes the other way, and an ice age occurs, most of the northern hemisphere will become uninhabitable.

We have been warned by scientists and other concerned citizens for many decades now about what we are facing, with little attempt being made to even listen to them, let alone take stock of what they are saying and react to the warnings. This is not a Nostradamus-type prediction, to be either taken or left, depending on whim or fancy. This is not a millions-to-one chance collision with an asteroid, so fascinating to movie goers. Nor is it the ultimate Armageddon of the fanatical religious sects. This is fact. This is happening.

"Is the surface of the Earth really the right place for an expanding technological civilisation?"

Dr. Gerard K. O'Neill*[17]

*Dr Gerard K. O'Neill was the principle proponent of extra-terrestrial habitation.

IS IT ALL OVER?

So, is it all over bar the shouting? I would say, not yet, but we will have to make radical changes to our expectations and our behaviour. The above quote sums up our predicament in a few short words - basically, that we have become too smart for our own backyard.

Not only was Professor O'Neill an outstanding physicist he was also a brilliant visionary when it came to the future of the human race. His book, *The High Frontier*, was an opening chapter on the colonisation of space. Where other writers, mostly science fiction, had proposed man's colonisation of the Moon, Mars and all points beyond for many decades, Dr O'Neill set down principles, discussed orbital space station designs, the physics of achieving this - first in 1974, barely six years after Neil Armstrong first set foot on the Moon, in a paper called The Colonisation of Space, following it later with his best seller, *The High Frontier*.

Dr O'Neill was of the opinion that such ventures were too important to the future of the human race to be left in the hands of national governments and he advocated that small groups of people should develop the tools of space exploration independently of governments, to prove that private groups could get things done enormously cheaper and quicker than government bureaucracies.

Six months after the publication of Dr O'Neill's book, on June 20, 1977, Anglia TV in England broadcast a documentary titled Alternative 3. The producers had intended to air a show about the British 'brain-drain' - about how British scientists were leaving the country to find higher-paying jobs abroad. But in the course of their investigations they discovered that many of the scientists they sought to interview weren't just leaving the country. They appeared to be mysteriously disappearing from the face of the Earth itself.

The disappearance of these scientists prompted the news team to investigate further, and what they claimed to have ultimately uncovered was a vast, global conspiracy reaching to the very highest levels of the American and Soviet governments. Apparently back in the 1950s researchers had learned that the Earth, on account of man's actions, was facing an unstoppable environmental catastrophe which would result in the almost certain extinction of humanity itself. World governments were left with only three options. They were:

Alternative 1 — to drastically reduce the human population on Earth;

Alternative 2 — to construct vast underground shelters to house government officials until the crisis had stabilised;

and

Alternative 3 — to establish a 'Noah's Ark' colony of humanity's best and brightest off of the planet, preferably on Mars.

The stony-faced earnestness of the announcer apparently convinced many that the documentary was real because after the show's conclusion Anglia TV was flooded with calls. If viewers had watched a little more closely, however, they would have seen that the copyright notice for the show was dated April 1 — April Fool's Day (even though the show wasn't broadcast on April 1st).

A few were so convinced by the show that they refused to believe it wasn't real, even after its producers announced that the entire thing had been a joke. These faithful few continue to insist that Alternative 3 is real, and that the show was part of the world government's vast and sinister disinformation scheme. They argue that by making Alternative 3 appear to be a fanciful hoax, the world government has insured that no one will suspect that it is, in fact, the frightening truth.*

---

* From www.museumofhoaxes.com

In hindsight, we can deduce that what was presented, and subsequently written in a book, also called Alternative 3, was a hoax. But what prompted it? First, there is the suggestion, from Dr O'Neill's book, that governments can't be trusted (so what's new?). Secondly, that already by that time scientists were expressing concerns over what we, the human race, were doing to this planet. And third, that speculations as to how the human race can survive a disaster as enormous as the destruction of our planet, were already rife in the early seventies.

We'll come back to this however.

The biggest stockpile of hydrocarbon fuel is not here on Earth. Remember that the universe is full of hydrogen. During the accretion of materials that would become our solar system, much of it was gas and as the gas and particles began to rotate the gas was not drawn into the core as much as the more solid material. The result was a string of planets, with solid planets closer to the Sun, gaseous planets on the outside - Mercury, Venus, Earth and Mars, are solid, Jupiter and Saturn are gas giants, while Uranus and Neptune are agglomerations of rock and ice, and helium - Pluto is a mongrel having probably been captured, rather than spawned by our Sun. Jupiter and Saturn are mostly hydrogen, with traces of other gases. What is interesting is speculation that some of the moons of Jupiter and Saturn might conceivably be covered with seas of liquid methane, ethane or butane. If only we could capture one of them, our problems would be solved! Is that not true?

Stop for a minute however, and think of the magnitude of the task, of even harvesting such fuel, should it exist. The distance to the nearest moon of Jupiter would be approximately the same as the distance to Jupiter at its closest approach to Earth, or nearly 4 AU*. That's, give or take a few metres, about 600,000,000 kilometres from Earth, or 1,200,000,000 km round trip.

---

*An AU, or Astronomical Unit, a measure of the distance from Earth to the Sun, roughly 150,000,000 km

Assuming we can design and develop failsafe automatic mining equipment (the climate in such places would be too harsh for even the bravest and most foolhardy of the human species to survive in, e.g., the intrepid drillers in the movie *Armageddon*), given today's propulsion systems, it would take many years to get there. Take your pick: liquid, as in the Saturn V Apollo missions - completely impractical, enormously expensive and very wasteful as a fuel in space, (such endeavours would have to be launched from space. Such equipment as needed to mine a planet's moon would be complex, thermally protected and simply too big to launch from an Earth station).

Another alternative would be solid fuel, a chemical plus oxidiser. The problem with solid fuel however would be you would need to carry the engine, to get you back, with you on the package, adding to the weight to be accelerated in the first place, as solid fuel rockets, once ignited are very hard to extinguish!

There are other alternatives, such as solar sails, nuclear rockets and other esoteric propulsion methods, none of which have gotten much further than the drawing board. Even Dr O'Neill's idea of a magnetic accelerator, or mass driver to ship materials back from the Moon has never gotten further than the laboratory, except perhaps as the basic fundamental behind maglev* trains.

But back to our quest. The translunar speed of the Apollo spacecraft was in the order of 40,000 kph. If we could devise a method to accelerate a mining system out into space towards a Jovian moon at that speed, it would take about 20 months just to get there. And that's in a straight line! Space trajectories are curved (because of the rotation of the planets) so it would in fact take considerably longer than 20 months to get there. After mining (the system would have to be automatic, it would take too long for signals to be received and answered from Earth), it would take another 20 months to get a loaded space tanker back here.

---

*Maglev, or magnetic levitation is the driving force behind such trains as the Japanese high speed test train that reached 550 km/hr

Going back to a previous chapter, if you remember, we are consuming on average about 85 million barrels oil equivalent A DAY. Even assuming the mining equipment is in place and the transporter ready to go, it would need to be bringing with it something like 50 BILLION barrels oil equivalent, just to tide us over until the next transporter arrived. Ah, but we could have many space tankers, just as we have many sea-borne tankers today. Okay, let's say we send one a month. That's still 2.5 BILLION barrels of liquid methane heading in our direction. We are very savvy now when it comes to space travel, we can control things remotely, even sending a washing machine-sized probe crashing into the backside of a comet. But what if, just supposing what if, one of those things (just imagine how big a 2.5 billion barrel tanker would be if we could build it) missed Earth orbit, and just sailed on into the middle of Times Square, or Piccadilly Circus? You go figure. Ten barrels is about 1.6 cubic metres. That's about 640 million cubic metres. I'm not even going to work it out, it is so absurd.

So here we face a problem. Mining distant planets, asteroids, moons for material to further the existence of man on Earth is not a viable proposition. The logistics of moving stuff here would defeat us. We consume too much to replenish our needs from space. One final comment - it cost billions of US dollars to bring back a few hundred kilos of Moon rock. How much would it cost to bring back even 1000 tonnes of Jovian moon?

It would seem therefore that our existence on Earth as we are right now, is a terminal prospect, except for the existence as previously described. However, Dr O'Neill in The High Frontier proposed several space habitats that could possibly work. Around the Earth-Moon system, there are several regions called Lagrange Points, named after an Italian-French mathematician Josef Lagrange, who discovered five special points in the vicinity of two orbiting masses where a third, smaller mass can orbit at a fixed distance from the larger masses. This means that there are five points around our Earth-Moon partnership that, should something be placed there, it would remain there.

Any satellite placed in orbit around a larger body in space will tend either to fall in towards that larger body (like SkyLab) or shoot off at a tangent into space. Our Moon is forever falling in towards Earth, just as Earth is forever falling in towards the Sun. It is only the relative velocities of each that keep them, the Moon round the Earth, and the Earth round the Sun, in orbit. So any satellite placed in orbit around Earth, would need constant boosts to keep it in high enough orbit it wouldn't come crashing into the Earth, but not so much that it might decide to just up and leave. Lagrange discovered these five positions where the gravitational attraction of the Earth and Moon balanced each other out in such a way that any object (such as a space station) placed there, would stay there - indefinitely.

Well and good so far. We have the technology now to build something like that, at one of the Lagrange points, just as we are building the International Space Station, assuming that we can build more reliable space rockets to get the material up there to begin with. One alternative would be to set up manned facilities on the Moon to process Moon rock into the building materials for a space habitat. All of this is possible, it is within our means, it is within our capabilities.

But what would we end up with? Even the largest space habitat imagined so far could not accommodate more than a few thousand, or tens of thousands of people. How to accommodate the 6 billion of today, the 10 billion people there will be by 2050? It's a dream, just as is any other alternative to survival.

Mind you, when you look at what I have described, and what was defined as the three alternatives, we're not really all that far off the mark. We know that we are systematically destroying this planet that we live on. Natural disasters, such as earthquakes, (not of our own doing), hurricanes (aided and abetted by our industrial outpourings?), floods and viruses (both natural and of our own doing), and bombings and conflicts (both of our own doing) are all helping to reduce the Earth's population. Maybe not as fast as they implied in Alternative 1. But that will be redressed soon enough as we begin to run out of readily available fuel.

Building underground caves for the chosen few to hide in as in Alternative 2? What about the nuclear bunkers into which presidents and prime ministers and others of their ilk would disappear, in the event of nuclear war? I really don't see too much difference.

And Alternative 3? If we built a space colony at each Lagrange point, who do you think would get to inhabit such places? Do they draw lots? Who has the power, the position to decide, anyway? Not that being chosen to inhabit one of these tin cans would appeal to me very much. Imagine spending your entire life in a crowded habitat, with thousands of other people - reminds me of the inner city slums which are, even now, already part of our lives. Not only that, imagine never being able to feel the warmth of the sun, bask in its golden glow, because you would have to be contained in heavy shielding to protect you from the energy waves coming from that raging nuclear reaction. There would be no atmosphere to protect you. I like the sea, the mountains, the open sky too much to ever want to live like that.

"At the end of a lecture on astronomy

given by a well known scientist (some say it was Bertrand Russell), a little old lady got up and told him he was talking rubbish, that everyone knew the world was a flat plate supported by a giant tortoise…"

from Stephen Hawking in A Brief History of Time[18]

12

ONLY TIME WILL TELL

Probably the most brilliant theoretical physicist since Albert Einstein, Stephen Hawking went on to say that most people would find the image of a universe as an infinite tower of tortoises rather ridiculous, but why should we think that we know better than the old lady? What do we really know about the universe, and how do we know what we know? Will we ever find the answers? Only Time, whatever that is, will tell.

And there, he has a point. I'm not going to delve into the space-time continuum or try and explain how new universes are created, but what I will tell you is that the Universe, the only one we know of right now, is ancient beyond imagination. It is about 13 billion years old - we know that by measuring the light from far distant objects.

We, on the other hand, have been around for about two million years. There is this enormously vast Universe just waiting for us to populate. It's waited many billions of years, it can wait a few billion more. The question is, will we ever populate even a small part of the Universe? As Stephen Hawking said, only Time, whatever that is, will tell.

One of the reasons I wrote this book was because we humans have a very self-centred, egocentric approach to everything. I wanted people to realise that this, 'Me first...' attitude will eventually be the end of the human race. We have to start thinking about the continuation of our race. Not ourselves individually, but the race as an evolving life form. It is in our genes to propagate the race. We will be faced with many decisions in the future, such as, if the habitats as described in the previous chapter are finally built, who will be chosen to go? Who will be left behind? Will choice be on intelligence? Looks? Financial worth? Capabilities?

Big decisions. But eventually they will have to be made. Will we travel to the stars? It's not beyond the bounds of possibility, but almost definitely not in our present form. If we assume that the rest of the planets in our solar system are uninhabitable, as we suspect, then our only alternative to survival will have to be towards the stars.

But survival takes on a whole new meaning when you think in those terms. We don't have the scientific knowledge to ensure the survival of the human race on an interstellar trip. The nearest star to us is more than four light years away, the distance that light, travelling at 300,000 km per second takes to reach it from here. So, travelling at the speed of light it would take more than four years to get there. We can't even approach one percent of the speed of light. Not even a hundredth of one percent of the speed of light, or 108,000 kph.

Except perhaps by one drive system, solar energy. The Sun emits a constant wind of particles that travel through space at enormous velocities. Collecting that energy in a solar sail, we could accelerate a space ship to speeds greater than anything we will ever achieve by any other method. But it will still take us thousands of years to get to the nearest star. We already know that it doesn't have a planetary system, so there is no point in heading in that direction anyway.

One final point. We look out into the Universe and we see stars. We see galaxies, billions of them, filled with billions of stars. We are looking at light from stars, from galaxies that might have exploded in super novae millions of years ago. It is little consolation to us to know that out there, assuming the laws of physics hold true throughout the Universe, there must be a myriad of other races, other sentient beings facing exactly the same dilemma as we are facing. The next stepping stone is just too far away for us to reach and the one we're standing on is becoming decidedly unstable.

I began this book with a quote from Dr David Price. It is perhaps fitting that I should therefore end with another quotation from the same source.

The human species may be seen as having evolved in the service of entropy, and it cannot be expected to outlast the dense accumulations of energy that have helped define its niche. Human beings like to believe they are in control of their destiny, but when the history of life on Earth is seen in perspective, the evolution of Homo sapiens is merely a transient episode that acts to redress the planet's energy balance. - Dr David Price[1]

Appendix

Appendix I

Petrochemicals and their products

Source	Level 1	Level 2	Level 3	Level 4
OIL & GAS	Ethylene	Ethylene dichloride, Polyethylene, Alpha olefins, Ethylene oxide, Ethanol	Vinyl chloride, Synthetic alcohols, Ethoxylates, Ethylene glycol ethers, Ethylene glycol, Ethanolamines, 1,3-propanediol	Polyvinyl chloride (PVC), Paints & coatings, Polyester polyols, Polyethylene teraphthalate, Antifreeze, Carterra
	Benzene	Ethyl Benzene, Cyclohexane, Cumene	Styrene, Adipic acid, Caprolactam, Phenol	Unsaturated polyester resins, Polystyrene, Acrylonitrile butadiene styrene (ABS), Styrene acrylonitrile (SAN), Styrene butadiene rubber (SBR), Styrene butadiene latex (SBL), Nylon 66, Nylon 6, Phenolic resins, Bisphenol A
	Propylene	Polypropylene, Acrylonitrile, Propylene oxide, Isopropyl alcohol, n0butanol/2-ethylhexanol	Acetone, Acrylic fibres, Adiponitrile / Hexamethylenediamine, Propylene glycol ethers, Propylene glycols, Polyether polyols	Methyl methacrylate, Direct solvent, Solvent Aldols, Polyurethanes
	Butadiene	Polybutadiene rubber		
	Toluene		Solvents, Toluene diisocyanate	
	Xylenes	Para-xylene	Teraphthalic acid	
	Methanol	Acetic acid, Formaldehyde	Vinyl acetate, Urea-formaldehyde resins	Polyvinyl acetate, Polyvinyl alcohol
	Butylenes	Methyl tertiary butyl ether, Secondary butyl alcohol, 1-Butene	Methyl ethyl ketone, Polybutylene	

Source: Shell Petrochemicals

Level 5	Products
	Bottles, food boxes, grocery bags, food wrapping, pipes, household detergents, industrial and institutional cleaning products, personal care products, plasticizers, lubricant additives, industrial surfactants, gas purification, coolants, powder and gel coatings, chemical intermediaries, pipes, cabling, windows/doors, packaging, clothing, flooring, insulation products, soft drink bottles and containers, polyester fibres (clothing, upholstery, carpets, pillows, surgical supplies, textiles, films, non-woven fabrics, engineered thermoplastics
Epoxy resins, Polycarbonate	Boats, jet skis, bathtubs, bowling balls, packaging, food & drink containers, automotive components, insulation construction materials, carpet backing, adhesives, asphalt, houseware, medical goods, tyres, shoe soles, conveyor belts, rubber tubing, tank and caterpillar tracks, sporting goods, toys, paper products, clothing, tyre cord, rope, parachutes, wood adhesives, insulation, laminates, coatings for wire and cans, computer circuit boards, composites, aerospace and marine components, windmill blades, glazing, optical media, autoparts, appliances, computers etc.
Polymethyl methacrylate	Food tubs, bottles, toys, storage boxes, battery cases, diapers, rope, clothing, geotextiles, food and flower wrapping, coatings, household products, pharmaceuticals, amines, solvents, ether derivatives, plasticizers, houseware, paints and coatings, cleaning solvents, electronic components, boats, bathtubs, water tanks, food, cosmetics, antifreeze, personal care products, adhesives, surfactants, rubber chemicals, mining chemicals, additives, flexible/rigid foam, mattresses, seating, automotive parts, sports tracks, playground surfaces, ski suits, light fixtures, medical devices, signs
	Tyres
	Paints, adhesives, printing inks, pharmaceuticals, degreasers
	Adhesives, insulation, automotive parts, electronics, textile finishing, films, paper products
	Gasoline, coatings, adhesives, pipes

Now let us do a little exercise in lateral thinking. First of all, how many of the products, on the right of the chart, do you think we could have manufactured, supposing we had alternative sources of energy, but no petrochemicals to make them from? Go through the list and see how many. Ten? Twenty? Remember, if you never had, you never wanted, so many of the products are additional benefits to our lifestyle that we would never have had if there were no petrochemicals.

But, work from the other direction now and consider the impact it is going to have on your life as all these products cease to exist or become too expensive to afford. Look at the middle of the list, there is one item there, windmill blades. Nothing special about them, is there? The Dutch had windmills hundreds of years ago, right? Wrong. These are highly specialised blades, for highly specialised windmills, the ones we use in wind farms to generate electricity from wind power. If we don't have the materials like resins, carbon fibre, laminates, etc., to make them from, we won't have the wind farms. Not without a great deal of investment in new materials technology. Hydrocarbon products are very strong and can withstand the stresses of something like being an atom in a piece of material at the end of a windmill blade, travelling at a speed approaching that of sound. Metals find such a situation almost suicidal. Propeller driven aircraft were not limited to the speeds at which they flew because of the airframe, they were limited by the tip speed of their propellers - too fast and they came apart. The same thing happens to a wind farm windmill, it can only go at a certain speed and no faster.

Appendix II

Installed geothermal generating capacities world-wide from 1995 to 2000 (from Huttrer, 2001), and at the end of 2003 (MWe - MegaWatts electricity equivalent energy)

Country	1995	2000	1995–2000		2003
	(MWe)	(MWe)	Increase (MWe)	Increase %	(MWe)
Argentina	0.67	-	-	-	-
Australia	0.15	0.15	-	-	0.15
Austria	-	-	-	-	1.25
China	28.78	29.17	0.39	1.35	28.18
Costa Rica	55	142.5	87.5	159	162.5
El Salvador	105	161	56	53.3	161
Ethiopia	-	7	7	-	7
France	4.2	4.2	-	-	15
Germany	-	-	-	-	0.23
Guatemala	-	33.4	33.4	-	29
Iceland	50	170	120	240	200
Indonesia	309.75	589.5	279.75	90.3	807
Italy	631.7	785	153.3	24.3	790.5
Japan	413.7	546.9	133.2	32.2	560.9
Kenya	45	45	-	-	121
Mexico	753	755	2	0.3	953
New Zealand	286	437	151	52.8	421.3
Nicaragua	70	70	-	-	77.5
PNG	-	-	-	-	6
Philippines	1227	1909	682	55.8	1931
Portugal	5	16	11	220	16
Russia	11	23	12	109	73
Thailand	0.3	0.3	-	-	0.3
Turkey	20.4	20.4	-	-	20.4
USA	2816.7	2228	-	-	2020
Total	6833.35	7972.5	1728.54	16.7	8402.21

Source: International Geothermal Association

Appendix III

Source / Effect	Wood	Coal	Oil	Gas	Nuclear	Geothermal	Direct Solar	Wind	Hydro & tidal
Produces H_2O vapour	X	X	X	X					
Produces CO_2	X	X	X	X					
Produces SO_2		X	X						
Produces CO	X	X	X	X					
Produces NO_x		X	X	X					
Particulate (airborne) matter	X	X	X						
Radioactive waste (residue)					X				
Chemical waste (residue)	X	X	X		X	X			
Water character changes					X				X
Consumes oxygen	X	X	X	X					
Lead emissions		X	X						
Hydrocarbon emission			X	X					
Adds to greenhouse effect	X	X	X	X	X	X			
Produces historical CO_2*		X	X	X					
Uses non-renewable source		X	X	X					
Basically no effect							X	X	

Note: Many more 'Effects' could be added and not all the sources produce the effects listed all of the time, but the basic idea can be seen from this table and that is that 'fossil' fuels, coal, oil and gas are the most damaging to our environment (the more 'X's earned, the more damaging). But we still consume them in ever increasing quantities - because they are the most versatile, the most readily available.

* 'Historical' CO_2 as opposed to 'normal' CO_2 is that which has been sequestered underground, in coal, in oil etc., for many millennia. Our atmosphere contains CO_2 which is cycled through our and other animals' respiration, the burning of wood and other biomass which is of recent origin. The atmosphere maintains a balanced CO_2 content. When we burn these fossil fuels, they release 'historical' CO_2 - a product that has been buried for millions of years, adding to the CO_2 content of the Earth's atmosphere. There can be little doubt, considering recent global weather and oceanic conditions, that adding this historical CO_2 is having a dangerous effect on our planet.

Appendix IV

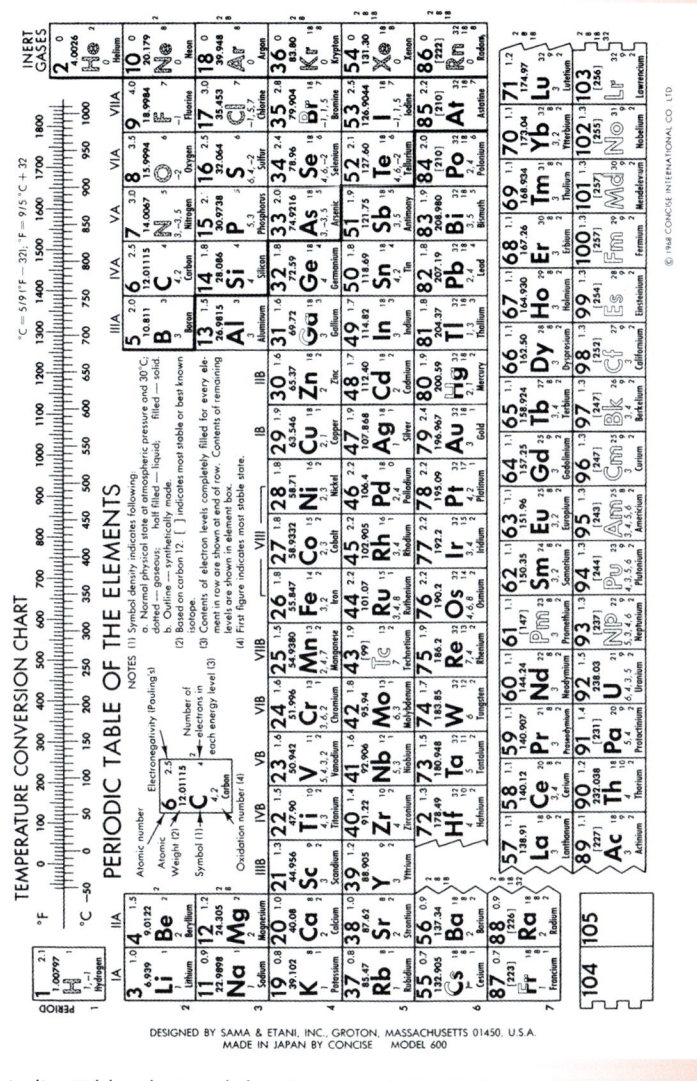

Periodic Table designed by Sama and Etani, Inc., of Mass. USA.
Unquestionably one of the best and most informative in the world with
regards to the natural elements.

Appendix V

Molar ionisation of selected elements vs. electronic configuration

Shell	K			L							M
	1s	2s				2p					3s
Z	1	2	3	4	5	6	7	8	9	10	11
1 H	1310										
2 He	2370	5200									
3 Li	520	7300	11800								
4 Be	900	1800	14800	21000							
5 B	800	2400	3700	2500	32800						
6 C	1090	2400	4600	6200	37800	47300					
7 N	1400	2900	4600	7500	9400	53300	64300				
8 O	1310	3400	5300	7500	11000	13300	71300	84100			
9 F	1680	3400	6000	8400	11000	15200	17900	92000	106100		
10 Ne	2080	4000	6100	9400	12200	15200				131200	
11 Na	500	4600	6900	9500	13400	16600	20100	25500	28900	141000	158700
12 Mg	740	1500	7700	10500	13600	18000	21700	25700	31600	35400	169900
13 Al	580	1800	2700	11600	14800	18400	23300	27500	31900	38500	42600
14 Si	790	1600	3200	4400	16100	19800	23800	29200	33900	38700	45900
15 P	1010	1900	2900	5000	6300	21300	25400	29800	35800	40900	46300
16 S	1000	2300	3400	4600	7000	8500	27100	31700	36600	43100	
17 Cl	1260	2300	3800	5200	6500	9300	11000	33600	38700	43900	51200
18 Ar	1520	2700	3900	5800	7200	8800	12000	13800	40800		
19 K	420	3100	4400	5900	8000	9600	11400	14900	17000	48600	
20 Ca	590	1100	4900	6500	8100	10500	12300	13800	18200	20400	57100
21 Sc	630	1200	2400	7100	8900	10700	13400	15300	17400	21800	24100
22 Ti	660	1300	2700	4200	9600	11600	13600	16600	18700	20900	25600
23 V	650	1400	2800	4700	6300	12400	14500	16800			
24 Cr	650	1600	3000	4800	7100	8700	15500	17800	20200		
25 Mn	720	1500	3300		7300		11500	19000	21400	24000	
26 Fe	760	1600	3000				14600	22600	25300	28000	
27 Co	760	1600	3200								
28 Ni	740	1800	3400								
29 Cu	750	2000	3600								
30 Zn	910	1700	3800								

This table is only to illustrate the amount of energy necessary to remove successive electrons from an atom. The values are irrelevant, they could be in ounces of gold, molar energy or bananas, it doesn't matter. What matters is the changes in energy levels from one electron to the next.

Look at No. 6, C, or Carbon. The first four values are, relatively speaking, quite low. The fifth and six values are six to eight times as high. This explains why Carbon can be free and easy with its four outer electrons and will share them with similarly minded atoms. Do you think that the fifth and sixth electrons participate in any reaction? Not likely is it, given the 'bribe' necessary to get the nucleus to let go of the electrons?

Conversely, it shows you how much energy there must be in a plasma, where all atoms are stripped of their electrons. A plasma is a prerequisite of a fusion reaction.

The Shell and Subshell labels simply indicate the grouping of the electrons, if you remember, into stable shells, or clouds mentioned earlier. Count them, 2, 8, etc - or check the energy levels from the Table of Elements. Only a few levels are shown here.

Appendix VI

It's all to do with transport!

Mode of transport	Purpose	Energy source	Prospects	Long term alternatives
Commercial aircraft	Local and international transportation of cargo and passengers	Kerosene (hydrocarbon fuel)	Terminal in the long run	Not in this lifetime
Commercial shipping	Local and international transportation of cargo	Bunker fuel, diesel, coal (hydrocarbon fuel)	Terminal in the long run	Nuclear for a period of time, wind power (sails).
Railroad	Local and long distance transportation of cargo and passengers	Diesel, electricity, coal (hydrocarbon fuel)	Limited future	Electricity while available
Trucks & road trains	Local and long distance haulage of cargo	Diesel (hydrocarbon fuel)	Terminal in the long run	None as yet
Passenger vehicles	Personal transport	Gasoline, NGV, electricity (mostly hydrocarbon, or hydrocarbon-based fuel)	Limited future	Electricity while available
Ox cart	Local haulage of cargo	Grass	Continue as before	Continue as before
Roller skates	Short distance travel	Food	Limited future	None

The purpose of this table is two-fold, the first to show you how dependent we are on hydrocarbons as our energy source for transportation. The list covers the five basic forms of transport over medium to long distances. It also shows the prospects for each in the long term. The second purpose is to show the alternatives once we no longer have hydrocarbons as an energy source.

Our race has expanded and grown through the use of transport to take people all over the world; to transport goods (trade); to distribute commodities. If you look at the graph on population vs. production elsewhere in this book, you will see that the population and hydrocarbon production match each other fairly well in exponential growth.

Hydrocarbon production will taper off rather rapidly over the next fifty years or so. What will happen to the population? You can't trade very far afield if the only source of transport you have is the ox cart, so where is industry to find its outlets? With no outlets for its production, industry will grind to a halt, people will lose their jobs and won't be able to afford the basic necessities, as food will not be available other than what can be produced locally. An ox can only haul a cart so far.

Our descendants, a few generations from now will have to relearn the skills of the agrarian ancestors we left behind a thousand years ago. What price a city view skyscraping condo when there is no electricity to run the elevators? Where will they get food from when the city is deserted? Life is going to be different. A lot different.

Additional sources of reference used in this book

o   *100 Years of the Pecten - Shell*

o   *Analysis of the Impact of High Oil Prices on the Global Economy -* IEA

o   Australian Cooperative Research Centre for Renewable Energy (ACRE) Ltd

o   Australian Research Institute of Renewable Energy

o   BBC News

o   CERA Advisory Service

o   Climate Change

o   *Coal, Natural Gas, and Petroleum*: Energy in the American Economy

o   Douglas-Westwood Ltd

o   *Economic Theory And A Faith-Based Approach To Global Warming* - Ferdinand E. Banks

o   EIA Annual Energy Review

o   EIA World Oil Supply

o   Energy & Geoscience Institute at the University of Utah

o   Energy Charter Secretariat - Annual Report

o   Energy Economics Council

o   *Energy Needs, Choices and Possibilities* - Shell

o   EPNews Service

o   ExxonMobil

o   *Fossil Fuels, Renewable Resources and Wind* - The Science Museum

o   Geohive

o   *Getting Down To Earth* - Joseph A. Tainter

o   IAEE

o   Idaho State University

o   IEEE

o   IPCC

o   Ms Cynthia Long

- o National Geographic magazine (especially June 2004)
- o New Scientist
- o Nuffield Advanced Science Book of Data
- o Oil Price Information Service
- o OPEC
- o Philip J. Grandinetti
- o ScienceNewsOnline
- o Scientific American
- o SEAPEX Press
- o Society of Petroleum Engineers
- o SouthEast Asia Petroleum Exploration Society
- o Standard Oil
- o Statistical Review of World Energy - BP
- o The Atlantic Report
- o The Economist
- o The Millennium Ecosystem Assessment
- o *The Realities Of The New International Energy Markets* - Terence H Thorn
- o The Science in Science Fiction
- o US Census Bureau
- o US Geological Survey
- o WebElements
- o *Whatever happened to Standard Oil?* - Chevron Corporation
- o www.chinaonline.com
- o www.rigzone.com
- o www.dieoff.com
- o Plus many, many - too numerous to count – other Websites

Endnotes

1   Dr David Price, Cornell University, Ithaca, NY 14853. *From Population and Environment: A Journal of Interdisciplinary Studies*, Volume 16, Number 4, March 1995, pp. 301-19 1995 Human Sciences Press, Inc.

2   Zayn Bilkadi, *Babylon to Baku*, Stanhope Seta Limited 1996

3   The United States DOE

4   *The History of Oil.*, Samuel T. Pees, Petroleum Geologist, Meadville, PA 16335 USA

5   *The History of The Standard Oil Company*, Ida M. Tarbell, 1904, McCLURE, PHILLIPS AND CO.

6   Michael Economides & Ronald Oligney, *The Color of Oil*, Round Oak Publishing, 2000

7   The Atlantic Report, H. D. Lloyd, March 1881

8   Berry Ritchie, *Portrait in Oil An illustrated history of BP*

9   Anthony Samson, *The Seven Sisters - The Great Oil Companies and The World They Shaped*, Viking Press, 1975; ISBN 0-553-13940-1

10 OPEC - The Organisation of Petroleum Exporting Countries

11 Joseph A. Tainter, *Complexity, Problem Solving and Sustainable Societies*, 1996, from *GETTING DOWN TO EARTH: Practical Applications of Ecological Economics*, Island Press, 1996; ISBN 1-55963-503-7

12   *Rutherford - a brief biography* - John Campbell, Canterbury, NZ

13   The Economist, August 20, 2005 'Boom and bust at sea'

14   *The Seven Sisters* by Anthony Sampson, Viking Press 1975; The Prize by Daniel Yergin, Simon & Schuster 1991; *The Carbon War* by Jeremy Leggett, Penguin 1999; *The Color of Oil* by Michael Economides and Ronald Oligney, Round Oak Publishing 2000: *The End of the Oil Age* by Dale Allen Pfeiffer, Lulu 2004; *Out of Gas - The end of the Age of Oil* by David Goodstein, Norton 2005 plus many others.

15   http://www.hubbertpeak.com/; http://www.odac-info.org/; http://www.dieoff.org/; http://www.peakoil.net/; the list is endless.

16   Fred Hoyle, *Of Men and Galaxies*. 1964, University of Washington Press, Seattle, USA

17   Gerard K. O'Neill, *The High Frontier*, William Morrow, Jan. 1977

18   Stephen Hawking, *The Illustrated Brief History of Time*, Bantam Books, 1996

19   Fred Hoyle, *Of Men and Galaxies*. 1964, University of Washington Press, Seattle, USA